Environment and Societ

Ethiopia is facing environmental and poverty challenges and urgently needs effective management of its environmental resources. Much of the Ethiopian landscape has been significantly altered and reshaped by centuries of human activities, and three-quarters of the rural population are living on degraded land. Over the past two decades, the country has seen rapid economic and population growth and unparalleled land use change.

This book explores the challenges of sustaining the resource base while fuelling the economy and providing for a growing population that is greatly dependent on natural resources for income and livelihoods. Adopting a political ecology perspective, this book comprehensively examines human impacts on the environment in Ethiopia, defining the environment both in terms of the *quantity* and *quality* of renewable and non-renewable natural resources. With high levels of economic production and consumption also come unintended side effects: waste discharges, emissions of pollutants, and industrial effluents. These pollutants can degrade the quality of water, air, land, and forests as well as harm the health of people, animals, and other living organisms if untreated or disposed of improperly. This book demonstrates how the relationship between society and environment is inherently and delicately interwoven, providing an account of Ethiopia's current environment and natural resource base and future considerations for environmentally sustainable development.

Girma Kebbede is Professor of Geography and Environmental Studies at Mount Holyoke College, South Hadley, Massachusetts, USA.

Routledge Studies in Political Ecology

The *Routledge Studies in Political Ecology* series provides a forum for original, innovative, and vibrant research surrounding the diverse field of political ecology. This series promotes interdisciplinary scholarly work drawing on a wide range of subject areas, such as geography, anthropology, sociology, politics, and environmental history. Titles within the series reflect the wealth of research being undertaken within this diverse and exciting field.

Political Ecology of the State
The basis and the evolution of environmental statehood
Antonio Augusto Rossotto Ioris

Political Ecologies of Meat
Edited by Jody Emel and Harvey Neo

Political Ecology and Tourism
Edited by Sanjay Nepal and Jarkko Saarinen

Environment and Society in Ethiopia
Girma Kebbede

Environment and Society in Ethiopia

Girma Kebbede

Routledge
Taylor & Francis Group

LONDON AND NEW YORK

First published 2017 by Routledge

2 Park Square, Milton Park, Abingdon, Oxfordshire OX14 4RN
711 Third Avenue, New York, NY 10017

Routledge is an imprint of the Taylor & Francis Group, an informa business

First issued in paperback 2018

British Library Cataloguing in Publication Data
A catalogue record for this book is available from the British Library

Library of Congress Cataloging in Publication Data
A catalog record for this book has been requested

ISBN: 978-1-138-20669-4 (hbk)
ISBN: 978-1-138-32457-2 (pbk)

Typeset in Times New Roman
by diacriTech, Chennai

Contents

List of figures		vii
List of tables		ix
Preface and acknowledgements		xi
List of abbreviations		xiii
Administrative regions and zones of Ethiopia		xv
1	Introduction	1
2	Soil resources and land use	14
3	Pastoral resources	52
4	Forests and woodlands	78
5	Freshwater	112
6	Wildlife and protected areas	153
7	Conclusion	183
	Bibliography	188
	Index	210

Figures

0.1   Administrative map of Ethiopia   xv
5.1   Major river basins of Ethiopia   116
5.2   The GERD under construction: 40 per cent completed by
      March 2015   124
5.3   Lake Abijata is shrinking in size and depth. The foreground
      used to be part of the lake   132
5.4   Dwindling acacia forest fringing Lake Shala   132
5.5   Lake Hawassa: eutrophication in process   135
6.1   Awash National Park: Beisa Oryx   159
6.2   Nech Sar National Park: Burchell's zebras standing under
      dry acacia trees   169

Tables

2.1	Per cent of farm holders by size of holding (hectare) by states	20
2.2	Ongoing major irrigation projects (as of 2010)	25
2.3	Regional distribution of land for investment	27
6.1	National parks and protected areas in Ethiopia	156
6.2	Endangered animals in Ethiopia	166

Preface and acknowledgements

Natural resources form the mainstay of the Ethiopian economy. Rural communities, the homes of nearly 80 million people, are heavily dependent on nature's bounties for their food, energy, shelter, and income. Their welfare in both the short and long term is closely linked to the productivity and sustainability of the country's ecosystems that support life. This book deals with a series of environmental and natural resources issues in Ethiopia – ranging from soils and land use, forests and woodlands, rangelands, freshwater, and wildlife and protected areas. It provides a detailed account of how human activities have degraded, depleted, or reduced the productivity of these essential natural assets.

The study methodology combines extensive field investigations and travels in various agro-ecological regions of the country and the collection and synthesis of literature on Ethiopian natural resources, including unpublished government reports, censuses, proceedings of numerous conferences and workshops organised by government and non-government institutions, and professional organisations. Field surveys were conducted in several districts and zones of all the regional states except, Gambela and Somali. The primary methods of field inquiry included taking notes of what was being observed in the physical environment and human-environment interactions, conducting informal individual and group interviews with farmers, local community leaders, community organisations, and district officials.

Writing this book was possible through the contribution of many people with whom I have interacted during a series of fieldwork I conducted across the country and during information gathering at many government and non-government institutions. I would like to thank all the individuals who gave their time willingly and provided useful information. In particular, I appreciate district officials who assisted me with logistics and safe accommodations. I am also grateful to the staff of the Document Center at the Ministry of Water Resources in Addis Ababa for allowing unrestricted access to their collections on water resources and related subjects.

I owe a debt of gratitude to three institutions. Mount Holyoke College, my home institution, provided the sabbatical leave and faculty grants to carry out a series of fieldwork in Ethiopia. A 2014–2015 Fulbright Fellowship award to teach and research in Ethiopia allowed me to conduct further fieldwork and to complete a draft that has culminated in this volume. Debre Berhan University (DBU), which hosted me as a Fulbright Scholar, provided logistical assistance and a favourable social and intellectual environment. DBU-organised and funded visit to the Grand Ethiopian Renaissance Dam under construction was, indeed, a memorable event. The trip afforded the opportunity to study the possible environmental impact of this ambitious project and the extent of land use and land cover in western Ethiopia. An equally unforgettable event for me was a two-week field trip, involving 36 senior Geography and Environmental Studies majors and five instructors, to the Ethiopian Rift Valley, the Bale Mountains National Park, the Sanetti Plateau, the Sof Omar Caves, the Awash National Park, Dire Dawa, and the historic city of Harar. This most rewarding educational field trip provided the chance to observe the amazing physiographic diversity, rich cultures, and human imprints on the landscape in eastern Ethiopia. I am also grateful for the opportunity to travel across the awe-inspiring landscapes of North Shewa, North and South Wello, Southern Tigray, Wag Hemira, South Gondar, and East and West Gojjam (see Figure 0.1).

I would like to acknowledge my colleagues in the Department of Geography and Environmental Studies at DBU for their collegiality, friendship, advice, and support during my Fulbright year.

My thanks also go to Christina Nyren, Associate Project Manager at the publisher, for her help with copy editing and formatting the manuscript. I am grateful for her patience, caring, and sound editing. I would also like to thank the anonymous referees who read the manuscript and provided me with encouraging and constructive comments.

My greatest debt goes to my brother-in-law, Dawit Gebretsadik, for assisting me in carrying out field surveys in Sodo Zuria, Arba Minch Zuria, Konso, Burji Amaro, Wendo, and Hawassa districts and the Nech Sar National Park and the Abijata-Shala National Park. I particularly appreciate his boundless patience, guidance, advice, and skilled driving.

As ever, my greatest thanks are due to my family for their love and support and for enduring prolonged absences from home during the fieldwork for the book.

Finally, while I acknowledge the help and support of individuals, groups, and institutions, any errors and deficiencies in this work are my responsibility.

Girma Kebbede, South Hadley, Massachusetts

Abbreviations

AF	Afar
APF	African Parks Foundation
ADLI	Agricultural Development-Led Industrialization Strategy
AM	Amhara
AWWP	Al Wabra Wildlife Preservation
ATNESA	Animal Traction Network for Eastern and Southern Africa
AVA	Awash Valley Authority
BE	Benshangul-Gumuz
BOD	Biochemical oxygen demand
COD	Chemical oxygen demand
CSA	Central Statistical Authority
CRV	Central Rift Valley
CFSCDD	Community Forestry and Soil Conservation Development Department
DoP	Declaration of Principles
DO	Dissolved oxygen
ENSO	El Nino/Southern Oscillation (ENSO)
EHT	Ethiopia Heritage Trust
EPRDF	Ethiopian People's Revolutionary Democratic Front
EVDSA	Ethiopian Valleys Development Studies Authority
EWNSH	Ethiopian Wildlife and Natural History Society
EWCA	Ethiopian Wildlife Conservation Authority
EWCO	Ethiopian Wildlife Conservation Organization
GA	Gambela
GERD	Grand Ethiopian Renaissance Dam
GLASOD	Global Assessment of Soil Degradation (GLASOD)
GDP	Gross Domestic Product
GNP	Gross National Product
GW	Giga Watt
IPCC	Intergovernmental Panel on Climate Change
ICARDA	International Center for Agricultural Research in the Dry Areas
ICRISAT	International Crops Research Institute for the Semi-Arid Tropics
ILO	International Labor Organization

ILRI	International Livestock Research Institute
IWMI	International Water Management Institute
ITCZ	Inter-tropical Convergence Zone
JIRCA	Japan International Research Center for Agricultural Sciences
MoA	Ministry of Agriculture
MoWR	Ministry of Water Resources
NMSA	National Meteorological Services Agency
NBI	Nile Basin Initiative
OR	Oromiya
NFPA	National Forest Priority Areas
NABU	Nature and Biodiversity Conservation Union
NTFPs	Non-timber forest products
PFMGs	Participatory forest management groups
PAs	Peasant Associations
PGRC	Plant Genetic Resources Center/Ethiopia
REST	Relief Society of Tigray
SST	Sea Surface Temperature
SMNP	Semien Mountains National Park
SO	Somali
SN	Southern Nations
SNNP	Southern Nations, Nationalities, and Peoples' Region
SFCDD	State Forests Conservation Development Department
SDPRP	Sustainable Development and Poverty Reduction Program
TI	Tigray
UNESCO	United Nations Educational, Scientific and Cultural Organization
UNFCCC	United Nations Framework Convention on Climate Change
WBISPP	Woody Biomass Inventory and Strategic Planning Project
WFP	World Food Program

Administrative regions and zones of Ethiopia

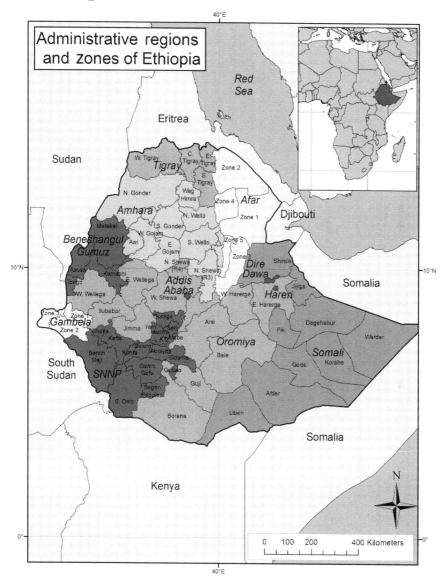

Figure 0.1 Administrative map of Ethiopia.

Source: Map produced by the GeoProcessing Lab at Mount Holyoke College.

1 Introduction

The economic and livelihood foundation of Ethiopia depends largely on primary resource extraction and export activities in agriculture. As a result, much of the Ethiopian landscape has been transformed by human activities. Some regions have been under the human influence for thousands of years, while others have only recently been altered by the expansion of agricultural, industrial, and mining activities. For Ethiopia to overcome widespread and extreme poverty, it needs sustained and sustainable economic development. Such efforts will require substantial increases in the use of its stock of natural capital: land, water, energy, pastures, forest and wildlife resources, and minerals. As citizens' standard of living improves, resource demands increase. Improved standards of living mean high levels of consumption and increased demand for food, water, energy, shelter, and other essential goods and services. In meeting these livelihood needs, people will have to draw on the country's natural resources.

Ethiopia is a country of extreme variation in the physical environment, natural resources, and cultures. It is one of the oldest political entities in the world, occupying a land area of a little over a million square kilometres – nearly the size of Texas and California combined. The country is in the high-sun zone, being situated between the latitudes three degrees north and 18 degrees north. In spite of its tropical location, a significant proportion of the county has a temperate climate due to high altitude. The high altitude not only reduces the occurrence of tropical disease but also causes increased precipitation. If it were not for its highlands, the country would likely just be part of the arid expanse of the Sahara Desert. Because of their volcanic origins, many areas of the highlands also have good soils. It is for these reasons that the highlands, while constituting only two-fifths of the country's area, are home to 80 per cent of the population and considered the nation's heartland.

Agriculture is the backbone of the country's economy, as measured by employment, contribution to total output, and export earnings. Four out of every five people in the country directly depend on farming and animal husbandry for employment, income, and food. Agriculture accounts for more than 45 per cent of gross domestic product (GDP) and 80 per cent of exports. Human settlements follow climate and productive soil. In highland areas where ample rain and rich soils are available, the human population is dense. The highland regions enjoy moderate climatic conditions: a cool, dry winter and a warm, wet summer. The vast semi-arid and arid lowlands areas of the country provide a livelihood to a significant number of pastoralists. The country has the highest number of cattle in Africa and is among the top 10 in the world. Over 50 million units of cattle are estimated to exist in the country, with large and small ruminants almost equal in number. It has substantial amounts of valuable mineral resources in the ground that have not yet been explored, including gold, platinum, copper, potash, nickel, and iron. Judging by its geologic history, the vast landscape of the Ogaden region possibly has substantial gas and oil reserves. The prospect of finding oil and gas in the Gambela regional states is quite high. The country is yet to fully exploit these mineral resources. The country's rivers that cascade down from its highlands have the potential to generate more than 45,000 MW of hydropower. In Sub-Saharan Africa, only the Democratic Republic of Congo tops this huge estimate. At present, it has developed only less than 10 per cent of this potential.

With a large geographic area, high agricultural potential, and substantial energy and mineral wealth, the country has high potential for sustained development. But in the past, this potential has not been sustainably utilised, largely because of prevailing extractive political systems and misguided government policies. The systems kept the populace under abject poverty and forced them to engage in ecologically harmful practices and natural resource extraction. In the pre-1974 revolution, the majority of the Ethiopian farmers worked under various exploitative land tenure systems and excessive taxation. Much of the products of their labour went to the landed aristocracy. The excessive expropriation of the farmers' produce prompted them to overuse the land, leading to soil degradation and erosion. Farmers were also disinclined to make investments in the land, such as erosion control or irrigation, as the benefit would go to the landlords and authorities. Inequities in the land tenure system also resulted in population pressure and caused more and more marginal land to be cultivated. Additionally, the exploitation of the farmers drove them to try to supplement their meagre incomes. Many adopted ecologically harmful practices, such as vegetation removal for sale. Such practices robbed the land of valuable nutrients and left it open to erosion. The lack of vegetation also reduced the land's ability to hold water and caused greater runoff, and thus more erosion. When fuelwood became depleted, farmers turned to using animal dung and crop residue for fuel, subtracting yet another source of nutrients from the land. The lack of entitlements caused poverty, which in turn caused destructive environmental practices.[1]

It is a common belief that agriculturalists and pastoralists are to blame for causing ecological degradation that ultimately undermines their livelihoods. However, this view ignores the economic, political, and social circumstances in which these systems operate. People knowingly do not impoverish the environment they depend on for their livelihoods. Whether natural resources are conserved or degraded depends largely on the existing economic, political, and social contexts.

Under oppressive economic conditions, farmers could hardly sustain themselves, let alone provide time and labour to develop the land. Any labour invested in planting trees, conserving soil or water, or building irrigation system would benefit almost exclusively the government and landlords in the form of rent and taxes. Also lacking was the impetus for technical creativity to increase production. In pre-1974 Ethiopia, there was a glaring contrast between the practices of farmers who did and did not own land. Farmers who had ownership title to a piece of land were much more interested in taking good care of the land they cultivated. They built terraces, dug irrigation channels, planted vegetation near the homestead, practised crop rotation, and added manure and other fertilisers to enrich the soil. Those who rented land to work on rarely took these measures.[2]

Under the repressive military regime, the 1970s and 1980s were a period of economic contraction rather than economic growth. The regime's imposition of state mandated centralised economic planning and the nationalisation of industries, urban properties, and rural lands set the stage for the deterioration of the Ethiopian economy. State ownership of productive enterprises resulted in production inefficiencies and the deterioration of public services. Misallocation of resources through price controls led to a retrenchment in economic activity. The land reform under the post-1974 government did not encourage farmers to consider their long-term impact on the land because they were not allowed to own the land. With no secure right to the land, most farmers had little incentive to invest in it. The state's top-down, bureaucratic approach undermined any use of local knowledge and farmers' participation. As a result, the support of the local farmers was undermined, and many of the environmental rehabilitation programmes failed to promote environmental stability. Worse, the state also appropriated any surplus through taxes, land use fees, and other compulsory contributions – further undercutting farmers' cooperation to participate in environmental restoration schemes. Predictably, most of the state-run environmental conservation and rehabilitation projects failed.

The Ethiopian economy at present bears little resemblance to the disastrous decades of the 1970s and 1980s. The 1990s and 2000s saw significant shifts from a highly centralised state- dominated economy to a market-oriented one that has resulted in sustained economic growth. The gross domestic product grew at an average of 5.6 per cent per year from 1992–2000[3] and at 8.8 per cent per year from 2000–2010,[4] more than six times than what it had been in the 1980s. The major thriving, growing sectors included agriculture, construction, real estate, and retail, followed by banking, insurance, and foreign investment. The private sector has grown dramatically. The country has invested huge sums of money in capital projects, such as roads and power plants. The economy is more diversified and less dependent on coffee for export. The country has made considerable investments in the social sector, particularly education, health, and infrastructure, which is having very significant impacts on economic growth.[5] It has also made progress toward meeting the Millennium Development Indicators, such as child mortality, maternal health, basic education, safe drinking water, and sanitation.

Under the current political system, ownership of land remains in the hands of the state. Under existing laws, only the state allocates land and selling land is prohibited. Farmers have user rights to the land with absolute title to the land vested in the state. But farmers have the right to transfer the land and the results of their investment to their offspring. Unlike the previous regime, the state does not dictate the crops farmers grow and the price they should receive for their produce. There is ample evidence that the existing tenure system and agricultural policies have improved the lives of the farming population of the country. Many have argued that public ownership of land removes incentives for land husbandry and long-term investment in land and resource conservation. But it appears that because farmers possess lifetime inheritable rights to the land they cultivate, they are more willing to make investments in and care for their land. In some regions of the country, farmers have doubled their grain production by adopting environmentally compatible technologies – including biological and physical soil and water conservation.[6] Institutional support services, such as input distribution, extension advice, and product marketing, were also significant factors for the relative improvement in productivity in the rural sector. The fact remains, however, ecologically incompatible land use practices remain destructive to the sustainability of the rural environment and natural resources. The key environmental issues in Ethiopia are soil erosion, rangeland degradation, deforestation, desertification, salinisation, siltation, water supply and shortages, water pollution, and loss of habitat and biodiversity.

Demographic trends

The impact of people on the use of renewable and non-renewable resources and the environment depends on their numbers and their level of consumption. The rapidly growing population of Ethiopia will certainly result in increasing land use alteration and exploitation of natural and environmental resources. The size and resources needed to sustain the additional members of the growing population will certainly affect the diversity of other organisms in the country as well. Ethiopia's population growth is among the fastest in the world. Up until the middle of the twentieth century, the population was growing at a slow rate. Although birth rates were high, they were nearly totally offset by high mortality rates – largely frequent famine, epidemics, and wars. Until 1900, the country's total population remained below 12 million inhabitants. Fifty years later, it had reached 19 million.[7] According to the nation's first census in 1984, Ethiopia's population was slightly over 42 million.[8] The 1994 census revealed that the population had grown to about 53 million.[9] Most recent (2016) figures show that the population had climbed to 96 million, making Ethiopia the second most populous country in Africa, behind only Nigeria,[10] and the 12th populous country in the world.

The population growth rate has been steady since 1960. The annual growth rate was estimated at 2.2 per cent for 1960–65 period, 2.3 per cent for 1965–70, 2.4 per cent for 1970–75, 2.6 per cent for 1975–80, and 2.8 for 1980–85.[11]

The 1990s saw a 2.9 per cent population growth rate. Estimated annual growth for the 2000s varied between 2.9 per cent and 3.2 per cent. This high growth rate is expected to continue for years because the country's birth rate (38.5 births per 1,000 in 2012) is still high.[12] By comparison, the death rate has fallen to 9.3 deaths per 1,000[13] because of a drastic reduction in child mortality. Over the last two decades, improvement in health services has helped newborns to have a greater chance of making it past their first year. As a result, infant mortality has declined from 111 deaths per 1,000 live births in 1990 to 68 in 2010.[14] Under-five mortality has also declined from 202 deaths per 1,000 live births to 106 deaths during the same period.[15]

The total fertility rate is also high, even though the rate declined from a high of 7.7 children per woman in 1984 to 3.9 in 2012, compared to 2.4 for the rest of the world.[16] Fertility rates are projected to continue declining, but not significantly enough to make a dent in the growing population for years to come. Moreover, population growth is likely to continue for several decades because a significant percentage of the population has yet to enter childbearing age. Like many other African countries, the Ethiopian population is disproportionately young. Currently, about 44.6 per cent of the population is under the age of fifteen,[17] compared to 26 in the rest of the world.[18] Thus, over the next ten to twenty years, these children and adolescents will reach their childbearing years and give momentum to the country's population growth. Even if the fertility rate is brought down close to the replacement rate, the sheer momentum of population growth means Ethiopia will keep growing dramatically in the coming decades, as those already born have children of their own. If current trends in population growth continue, there will be 129 million or more people by the year 2030 and 180 million or more by the middle of this century.[19]

A significant consequence of the high population growth is an increase in population densities. We consider here arithmetic density without regard for population distribution or the limits of arable land.[23] The overall, population density for the country jumped from 50 people per square kilometre in 1994 to 75 people per square kilometre in 2009[24] to 82 in 2015, compared to the world's population density of 49. If we exclude the arid and extremely cold and rugged mountainous regions, which are largely uninhabitable, the population density rises increases to 204 per square kilometre (assuming 40 per cent of the country can be habitable). If we consider the population in proportion to only arable land – land suitable for agriculture (totalling 154,602 square kilometres) – the density quickly rises to 582.[20]

The environmental and resource implications of this unprecedented population expansion are clear. As such, the current rate of population growth is unsustainable and should be seen as a serious concern. Over four-fifths of the Ethiopian population heavily relies on natural resource-based livelihoods, including agriculture, grazing, forest products, and fishing for income. This reality necessitates that natural resources and ecosystem services be used and managed sustainably. There is a strong relationship between population and consumption of environmental resources. As the country's population attains a better standard of living, per capita consumption of vital livelihood resources – such as food, water, wood, energy, and other goods – will grow accordingly. In meeting these needs, people draw on natural resources, some of which are already in short supply or are declining in quality. The capacities of existing natural resource bases to support additional population will become increasingly strained. To understand the challenges of natural resources conservation and management, we need to examine the state of the natural resource base of the country: land (soil), forests, rangelands, freshwater, and wildlife. These are renewable sources that can be regenerated. However, they can also be degraded or depleted as a result of over-exploitation and poor management practices. Environmental degradation intensifies especially when there is a high level of poverty, insecure rights to livelihood resources, and heavy dependence on natural resource extraction for subsistence.

The purpose of this book is largely to examine the human impacts on the environment in Ethiopia. The term 'environmental' refers to both the quantity and quality of renewable and non-renewable natural resources that people use to sustain their lives and the surrounding environment consisting of water, air, landscape, and the atmosphere. In this context, environmental degradation means the shrinking of the environment in terms of quantity and its depreciation in terms of quality.[21] Excessive land degradation, soil erosion, rapid deforestation, watershed degradation, loss of biological diversity, fuelwood shortages, urban congestion, and overgrazing are as common in Ethiopia as they are in other developing countries. Ethiopia urgently needs effective management of its environmental resources at least for two primary reasons. First, much of the Ethiopian landscape has been significantly altered and reshaped by centuries of human activities. Currently, nearly three-quarters of the rural population are living on degraded land.[22] Given the heavy ecological toll that the country has suffered over a long period of its history, environmental concerns should be on the forefront of the country's development strategy. Second, over the past two decades, the country has seen rapid economic and population growth and unparalleled land use changes in rural as well as urban landscapes. This reality reinforces the challenges of sustaining the resource base while fuelling the economy and providing for a growing population greatly dependent on the use and extraction of natural resources for their income and livelihoods. Growing economic activities, increasing population, and consumption levels necessarily entail the intensification of the use of natural and environmental resources: soil, water, air, forests, minerals, and other natural resources. With high levels of economic production and consumption come unintended side effects: waste discharges, emissions of pollutants, and industrial effluents. These pollutants can degrade the quality of water, air, land, and forests as well as harm the health of people, animals, and other living organisms if untreated or improperly disposed of.

This book is not all about doom and gloom. While it attempts to provide a realistic view of the state of the country's environment and natural resource base, it also provides some positive examples that provide hope for future environmentally sustainable development. The last two decades have seen increased efforts in the rehabilitation of degraded land and sustainable land management, including such efforts as afforestation, reforestation, agroforestry, area enclosure, water harvesting, zero grazing, protecting existing forests, and restoration of pasture-lands. Given the magnitude of land degradation in the country, these efforts have to be scaled-up manyfolds.

Ethiopia is facing environmental and poverty challenges. Poverty is still high in spite of the country's remarkable achievement in the rate of economic growth in the last two decades. In general, the poorer the household, the greater the reliance on the natural resource base; for example, firewood for cooking, fodder for feed-ing livestock, and wood for the construction of houses. When rural households are poor, they will be unable to invest in soil conservation and land improve-ment. Hence, the prospects for resource management in Ethiopia lies on efforts aimed at eradicating poverty through the improvement of the material well-being of citizens and sustainable development. Also, resources such as land, forests, rangeland, water, and wildlife should be managed in an integrated manner, not in isolation from each other, because of their interdependence. The future of the country's livelihood security will depend on how these resources are managed at present and in the future. Resource management should involve strategies and technologies for resource development to sustain economic development without causing needless environmental harm.

Hence, the prospects for natural resource management in Ethiopia lies on efforts aimed at eradicating poverty through the improvement of the material well-being of citizens and sustainable development. Sustainable development is inex-tricably linked with the protection of ecological goods and services that contribute to the human quality of life and the continued unimpaired existence of other living things. It means improving the quality of life while protecting nature and environ-mental resources. Sustainable development also includes elements of not harming the future. As the World Commission on Environment and Development noted, sustainable development is that which 'meets the needs of the present without compromising the ability of the future generations to satisfy their needs'. The Commission emphasised that sustainable development 'requires a change in the content of growth, to make it less material- and more energy-intensive and more equitable in its impact'.[23]

Organisation of the book

Land is the most important natural resource of Ethiopia, which is the foundation for the agricultural production and economic development. It is agriculture that feeds and sustains the vast majority of the population of the country. The sector is expected to continue feeding the growing population, to foster overall economic development, and provide employment. The challenges surrounding agriculture and land use in Ethiopia are numerous and varied. The soil is the most damaged ecosystem in the country, and its continuing degradation is extremely worrisome in terms of future agricultural production and productivity losses. Considerable quantities of topsoil are lost every year from croplands that produce most of the country's food. Ploughing the land year in and year out disturbs the topsoil and makes it easier for the soil to be eroded by rain and the wind. Plant nutrients (such as nitrogen, phosphorus, and potassium) and moisture are also lost along with the soil itself. The soil that is washed away from agricultural and non-agricultural land finds its way into water bodies. As a result, the lifetime of many of the country's water storage reservoirs has considerably been reduced by inputs of sediments from eroded croplands as well as from uncultivated lands without vegetation cover.

The country's arable land is under increasing pressure not only from a growing rural population, increasing urban demands but also from increasing conversion of land for the production of export-bound flowers, foreign-owned large-scale agricultural enterprises, and biofuels. One should not underestimate the direct adverse consequences of such activities on the environment. Large-scale mechanised farming activities will surely contribute to soil erosion, salinisation (accumulation of salt in the soil), and over-exploitation of water resources. One of the most direct effects of modern large-scale farming on the environment is contamination of surface waters and groundwater by fertilisers and pesticides. Excess nutrients have contributed to increases in the concentration of total dissolved solids and the eutrophication of the country's lakes and reservoirs.

Land use patterns vary enormously as a result of both climatic conditions and human activity. The variety of different agro-ecosystems in the country produces a wide range of crop diversity. Many of the country's crop species, such as sorghum, barley, tef, chickpeas, and coffee have several different landraces and wild varieties. Genetic diversity in crops allows farmers to take advantage of the many different microenvironments with their different soils, precipitation regimes, temperatures, altitudes, slopes, and fertility. Diversity also offers protection against diseases, pests, drought, and other natural stresses. Maintaining such genetic diversity is thus crucial to sustainable agricultural development in the country. However, this diversity is currently suffering from genetic erosion due mostly to the replacement of indigenous landraces by new cultivars, destruction of ecosystems, changes in agriculture and land use, loss of habitats, and climate change-induced factors, such as droughts.[24] Chapter 2 describes the physiographic and climate features and their impacts on livelihoods and land management, describes the important features of the different types of soils in Ethiopia and their suitability for agriculture, and discusses how soils are managed and mismanaged. The chapter also examines different agricultural systems and land use patterns and concludes with a discussion of major challenges to sustainable agricultural development.

Pastoralism – a way of life based on keeping herds of grazing animals – is an extensive form of land use in Ethiopia, and about approximately half of the country is estimated to be suitable only for animal husbandry. This form of land use typically occurs in the lowland, arid and semi-arid areas, mostly below 1,500 metres elevation. Precipitation is low and unpredictable and, compared to the highlands, the lowlands are at an increased risk of drought. As rain-fed agriculture is restricted, most of the population rely more on pastoralism. Pastoralists represent less than 10 per cent of the population, but they own more than a third of the country's cattle population. Pastoralism is the main source of livelihood for the vast majority of the people living in southern Oromiya, Somali, and Afar regional states. Small communities in southern SNNP, Gambela, and Benshangul-Gumuz regional states also subsist on pastoral and agro-pastoral forms of livelihood. Over the last half a century, pastoralists' rangelands have been shrinking as more and more land is converted to irrigation farmland and parks and is encroached by neighbouring communities. As pastoralists are pushed away from their traditional pastures, they are forced to induce overgrazing on the remaining land available for pasture. Traditionally, pastoralists have coped with drought by moving to areas with more forage and water. However, this movement is now constrained as settlers, whose numbers have grown due to population growth, have occupied many former grazing grounds. The degradation of the country's rangelands is not only detrimental to the livelihood of several million people but also exacerbates the adverse impact of climate change since range-lands capture considerable volumes of carbon in shrubs, bushes, and grasses. Chapter 3 discusses some of the physical and biological characteristics of the lowlands and the factors that may have been responsible for the degradation of pastoral resources. The chapter concludes that recurrent drought, displacement, inadequate feed, water shortage, livestock diseases, lack of adequate institutional support, and political disturbance and insecurity constitute major threats to the sustainability of pastoral livelihoods in lowland Ethiopia.

Ethiopia's natural vegetation is diverse. This diversity can be explained by prevailing environmental conditions, including rainfall, temperature, and soil. There is little doubt that forest cover in Ethiopia has decreased at a rapid rate, particularly in the northern highland regions of the country. Much of the country's original forest has been cleared or degraded. Fuel and construction needs, land for crop cultivation, and illegal timber harvesting are major factors for the diminishing of vegetation cover of these and other regions of the country. Today, Ethiopia has an estimated forest area of 13 million hectares, representing about 12 per cent of the country's total land. Forest areas in Ethiopia are declining by 1.1 per cent annually.[25] Major changes occur as a result of the removal of vegetation. Soil temperature rises and runoff and evaporation increase. If the land does not restore itself, loss of nutrients, soil erosion, and soil acidification may follow. Rain also penetrates less deeply into the ground and, as a result, the water table may drop. Hence, the country's agricultural productivity and biodiversity depend on the conservation, sound management, and sustainability of its forests and woodlands. Chapter 4 deals with the types of forest biomes in Ethiopia, their ecological importance, the extent and distribution of forests, and the scope and underlying causes of deforestation. It also examines why past forest conservation efforts have failed, and the lessons learned, if any, to protect the remaining forest ecosystems in the country.

The Ethiopian highlands, with the headwaters of the Blue Nile, Tekeze, Wabi Shebelle, and the Omo rivers, constitute one of Africa's huge storehouses of freshwater. Overall, annual rainfall is adequate, and the country has several major lakes and rivers and significant groundwater resources. Ethiopia's per capita share of the renewable water resource is better than most Sub-Sahara African countries. However, this situation is changing with increasing water demand induced by high population growth and increased socioeconomic development activities. The per capita availability of water has declined from 2,340 cubic metres in 1994 to 1,378 cubic metres in 2015. The decline is due to population growth rather than a decline in total water resources availability. A per capita availability of less than 1,700 cubic metres is considered a water-stressed condition.

Water resources in the country are unevenly distributed and accessed. Highlands have surplus water, while the vast lowland arid and semi-arid regions of the country often experience water shortages. Water shortages in the lowland are exacerbated by the country's erratic rainfall patterns and periodic droughts. One of the major features of the rainfall in Ethiopia is its unequal distribution during the year and its variation from year to year in terms of amount, occurrence, and duration.

Ethiopia has not managed to put its freshwater resources to good use nor has it been able to protect them from being degraded. Despite the overall availability of water in the country, water is not always readily accessible to its people. Only one in five households can access improved water within 500 metres of the home. Lack of access to adequate safe drinking water is one of the main factors why waterborne diseases are major causes of morbidity and mortality. The hydropower potential of the country alone is enough to provide many times the power currently consumed by neighbouring Sudan, Eritrea, Djibouti, Somalia, and Kenya combined. However, over 80 per cent of its people lack access to electricity, and 90 per cent of them live in rural communities.

Water resources are also poorly managed. Extensive ecological degradation has resulted in severe problems in terms of water conservation and maintaining the natural flow of rivers. As watershed areas without vegetation cover increase, water infiltration decreases, flooding and siltation increase, and water quality and quantity decline. Destructive and inefficient uses of water abound. Inefficient irrigation systems overuse water, resulting in increased salinisation, as well as the rising of the water table. There is also evidence that human activity has already degraded some of the country's aquatic environments. Toxic chemicals released by industries and municipal and agricultural wastes pollute water resources. Polluted water adversely affects the health of the people consuming it, causing reduced economic productivity, social suffering, and premature mortality. Water pollution also reduces the availability of freshwater for other purposes, absent treatment. In particular, watersheds near urban areas are prone to contamination with pollutants. The extraction of excessive amounts of water is also undermining the renewability that sustains the country's lake ecosystems.

Aquatic ecosystems are suffering from siltation, nutrient eutrophication, water loss, and pollution. Chapter 5 deals with water resource distribution, threats to water ecosystems, and water management and development problems and prospects in the country.

Chapter 6 assesses the state of national parks and protected areas in the country. Ethiopia is rich in biological diversity. It is home to many mammal species – 277 of them, of which 35 are found exclusively in Ethiopia. Ethiopia is also very rich in avifauna diversity, with no less than 800 bird species, 16 of which are endemic. There are as many as 8,000 plant species in the country, perhaps even more; most have yet to be identified, classified, or studied. The country has eighteen national parks, four wildlife sanctuaries, eleven wildlife reserves, and eighteen controlled hunting areas to protect wildlife. In all, these areas cover about 16.4 per cent of the total land area of the country, compared to about 11.5 per cent of the Earth's total land surface that is categorised as protected areas.[26] In almost all cases, however, the establishment of national parks and protected areas involved the exclusion of people living on the land. Inside or adjacent to all the national parks live tens of thousands of poor people who depend on the parks to provide a significant proportion of their livelihood needs. However, these people are excluded from accessing resources in the parks or benefiting from park revenues or having a say on park management. The result is that conflicts occur between the government that tries to protect wildlife and people whose livelihood relies on accessing resources in protected areas. In many areas, wildlife habitats are converted into farms and settlements. Commercial land-use activities, such as cash crop production, logging, charcoal production, and mining, have caused habitat and species losses. The government stresses the importance of Ethiopia's biodiversity to economic development, food security, and livelihoods. It also emphasises the importance of conserving biodiversity. In practice, however, local communities have little or no involvement in designing and managing protected areas. The contention of this chapter is that national parks and other protected areas are neither protected nor serve directly the needs of the local communities. As the population continues to grow, local people will have no other options other than overrunning the country's fragile ecosystems and the habitats of already depleted, threatened, or endangered species. The chapter underscores the urgent need for a community-based approach to conservation and effective management of wildlife resources. It argues that the sustainability of parks and protected areas is largely dependent on the extent to which local communities are allowed to become partners in the protection schemes, attain ownership of the schemes, and obtain economic benefits from them. For instance, revenues obtained from tourism should be either distributed to local communities or be invested in projects serving community interests. Many case studies of parks and other protected areas have proven that prohibiting all human use of protected areas has often been ineffective. Many conservationists now believe that human resource use must be incorporated in protected areas management strategies.

Notes

1 Girma Kebbede. 1992. *The State and Development in Ethiopia*. London: Humanities Press, p. 72.
2 Ibid., p. 71.
3 Alemayehu Geda. 2009. The political economy of growth in Ethiopia. In: *The Political Economy of Growth in Africa*: 1960–2000, Benno J. Ndulu, Stephen A. O'Connell, Robert H. Bates, et al. (eds.), Cambridge, UK: Cambridge University Press pp. 116–42; p. 119.
4 Armin Rosen. 2012. The Zenawi paradox: the Ethiopian leader's good and terrible legacy. *The Atlantic*, July 21. Available online at www.theatlantic.com/international/archive/2012/07/the-zenawi-paradox-an-ethiopian-leaders-good-and-terrible-legacy/260099/ (accessed 21 July 2012).
5 Stefan Dercon. 2010. Globalization and marginalization in Africa: poverty, risk, and vulnerability in rural Ethiopia. In: *The Poor under Globalization in Asia, Latin America, and Africa*, Mchinko Nissanke and Erik Thorbecke (eds.), Oxford: Oxford University Press, pp. 368–97; p. 392.
6 Third World Network. 2007. *The Tigray Experience: A Success Story in Sustainable Agriculture*, Geneva: Third World Network; cited in United Nations Conference and Trade and Development (UNCTAD). 2012. *Economic Development in Africa: Report 2012: Structural Transformation and Sustainable Development*, New York and Geneva: UNCTAD, p. 123.
7 Central Statistical Agency (CSA). 1984. *Ethiopia 1984: Population and Housing Preliminary Report*, 1, 1, Addis Ababa: CSA.
8 CSA. 1995. *The 1994 Population and Housing Census*, Addis Ababa: CSA.
9 CSA. 2010. *Statistical Abstract 2009*, Addis Ababa, pp. 21–48.
10 Helmut Kloos and Aynalem Adugna. 1989. The Ethiopian population: growth and distribution. *The Geographical Journal*, 155, 1: 33–51; p. 36.
11 US Central Intelligence Agency, The World Factbook. Available online at www.cia.gov/library/publications/the-world-factbook/geos/et.html (accessed 21 March 2013).
12 Ibid.
13 UNICEF, Ethiopia: basic indicators. Available online at www.unicef.org/infobycountry/ethiopia_statistics.html (accessed 21 March 2013).
14 Ibid.
15 Ministry of Health. 2005. *National Strategy for Child Survival in Ethiopia*, Addis Ababa: Family Health Department, Federal Ministry of Health. Available online at www.eshe.org.et/childsurvival/Child%20Survival%20Strategy.pdf (accessed 1 August 2011).
16 US Central Intelligence Agency, The World Factbook. Available online at www.cia.gov/library/publications/the-world-factbook/geos/et.html (accessed 21 March 2013); United Nations Development Program (UNDP). 2013. The rise of the South: Human progress in a diverse world. *Human Development Report 2013*, New York: UN, p. 196.
17 United States Population Reference Bureau. 2011. *World Population Data Sheet*. Available online at www.prb.org/DataFinder/Geography/Data.aspx?loc=278 (accessed 21 March 2013).
18 Getachew Minas. 2001. Population dynamics and their impact on population growth and sustainable livelihood in Ethiopia. In: Food Security through Sustainable Land Use, *Proceedings of the Second National Workshop of NOVIB Partners on Sustainable Land Use*, Taye Assefa (ed.), Addis Ababa, pp. 89–132; p. 91.
19 Greg Mills. 2010. *Why Africa Is Poor and What Africans Can Do About It*, Johannesburg: Penguin, p. 170.
20 Available online at http://en.worldstat.info/Asia/Ethiopia/Land (accessed 18 March 2014).
21 Theodore Panayotou. 1993. *Green Markets: The Economics of Sustainable Development*, San Francisco: Institute for Contemporary Studies, p. 3.
22 United Nations Development Program. 2013. *Human Development Report 2013*, New York: UNDP, Table 13, p. 192.

23 World Commission on Environment and Development. 1987. *Our Common Future*. Oxford: Oxford University Press, p. 52.
24 Melaku Worede, Tesfaye Tesemma, and Regassa Feyissa. 1999. Ch. 6: Keeping diversity alive: an Ethiopian perspective. In: *Genes in the Field: On-Farm Conservation of Crop Diversity*, Stephen B. Brush (ed.), Boca Raton, FL: CRC Press, pp. 143–61.
25 Tony Binns, Alan Dixon, and Etienne Nel. 2012. *Africa: Diversity and Development*, New York, Routledge, p. 94.
26 Martha Honey. 2008. *Ecotourism and Sustainable Development*, London: Island Press, p. 13.

2   Soil resources and land use

Ethiopia has had centuries of successful agriculture, predating the 2,000-year-old Aksum civilisation. Small farming communities dotted much of the landscape of northern Ethiopia about 4,000 years ago.[1] Well-established systems of cultivation involving the use of oxen-drawn ploughs, 'wood-hafted sickles,' and the practices of 'piedmont flow irrigation from cisterns' and crop rotation were seen 2,000 years ago.[2] Increased agricultural productivity yielded an increase in population and, in turn, induced the cultivation of more land. As a consequence, human activities have profoundly transformed the landscape of much of Ethiopia's highlands. Woody vegetation, which once covered much of the highlands, has been cleared for agriculture and grazing. Mountain forests have been relentlessly removed for fuel and construction. The loss of vegetation cover and continuous cultivation of the land over the centuries have led to the diminishment and deterioration of the fertile soil.

This chapter has three aims. First, it will describe the salient characteristics of the different types of soils in Ethiopia, consider what makes them suitable for different agricultural purposes, and discuss how farmers manage them. Second, it will examine different farming systems and associated problems. Third, it discusses the major challenges to sustainable agricultural development.

Major soil types

The soil is the productive top layer of the earth's crust. It is one of the most important elements of the land resources. Humans depend on the soil, like all plant and animal species. It is also the most vulnerable natural resource upon which humans have had indelible impacts, both beneficial and detrimental.[3] The formation of mature, life-giving soil is an intricate process that may require several thousand years. Several factors influence the genesis and the distinctive nature of the soil, including the type of geological material from which it originated, climate, living organisms, topography, how long the soil has been weathered,[4] and the artificial changes caused by human intervention.[5]

The characteristics of the soil, along with climate, topography, and vegetation, determine the habitability of a place for humans. Soils influence agricultural land uses. The Ethiopian landscape exhibits a diverse array of soil types. (Most end in *sol*, derived from the Latin *solum*, meaning soil). The National Atlas of Ethiopia identifies 17 different types of soil. The dominant ones include Nitosols, Leptisols, Lithosols, Cambisols, Vertisols, Calcisols, Luvisols, Gypsisols, Solochaks, Regosols, Fluvisols, Alisols, Aridisols, and Andosols.[6] Features such as parent materials, horizon patterns, texture, and chemical content characterise each soil.

Nitosols are deep, clayey red soils found mostly in areas receiving high to moderate amounts of rainfall, including parts of Metekel, Asosa, West Gojjam, East Gojjam, South Wollo, East Wollega, West Wollega, Jimma, Kaffa, Shaka, and Bench Maji zones.[7] Nitosols make up about 12.5 per cent of the soils found in Ethiopia and 23 per cent of the arable land. They can be found on either steep or flat terrain, where they are associated with vertisols or gleysols. The characteristics of nitosols differ depending on topography. On steep slopes, they are shallow and in the form of acrisols/luvisols and cambisols. In areas of high precipitation, dystric and humic nitosols can be found. These soils are excellent for growing coffee. Humic nitosols are located in the highlands and lowlands and are usually covered with natural vegetation. Areas with good drainage, such as transition zones from grassland to forest, usually have nitosols. Much of these lands, being well drained, also make for good agricultural land.[8]

Nitosols have a stable structure and are porous, making them easy to drain and work. They can store nutrients and water; therefore, their good chemical and physical characteristics, along with an occurrence in favourable climates, make them well suited for agriculture. In areas of moderate to high population, such as the central highlands, they can support both annual and perennial crops. In areas of lower population densities, like western Ethiopia, they are used to grow perennial crops like coffee and tea. As with other soils, nitrogen and phosphorus are the most likely nutrients to be deficient in nitosols. They leach when continuously farmed and so must be treated carefully.[9]

Cambisols: Dystric and humic cambisols makeup 9.5 per cent of the soils of Ethiopia and 19 per cent of the arable land. The steep slopes in Western and Central Tigray zones, North Shewa, parts of the Bale zone, East Hararghe Zone, and semi-arid parts in Afdem zone and Deghabur zone in Somali State have cambisols.[10] These soils usually occur in association with dystric nitosols and acrisols. Intensive cultivation over the centuries has eroded them extensively from the northern highlands, but they can still be found in colluvial materials on steep slopes, where they usually occur in association with lithosols. On flat land, they are associated with vertic luvisols and vertic cambisols. In the semi-arid east, southeast, and south, they are found on sloping to steep land in association with vertisols, entisols, and inceptisols. In eastern Ethiopia especially, they can also be found with lithosols in limestone areas.[11]

Vertisols are one of the most important and widely distributed agricultural soils in Ethiopia. They cover 10 per cent of the surface of the country or approximately 13 million hectares of land. Vertisols constitute about 18 per cent of the country's arable land and are the principal soils used for crop production. Vertisols typically occur on flat or undulating terrain; they are common in the central and eastern highlands, in eastern Hararghe, on the flood plains of the Wabi Shebelle and Fafan Rivers, in valleys in Tigray, and western Gambela. Highlands with elevations of more than 1,500 metres above sea level have about 7.6 million hectares of vertisols.[12] Vertisols are high in clay content and, thus, have high moisture retention. Compared to other soils, the fertility of vertisols is above average. However, vertisols often have deficiencies in nitrogen and phosphorus. Most vertisols are slightly acidic; 61 per cent are in the 5.5 to 6.7 pH range, 21 per cent in the 6.7 to 7.3 pH range, and 9 per cent have a pH greater than 9.[13]

These dark clay soils occur in areas where seasonal rainfall is high, sometimes exceeding 2,000 millimetres per year. They are mostly impermeable and can become waterlogged in response to heavy rains. They are hard to drain because their clay content is almost 50 per cent. It limits agriculture to a degree, as water does not infiltrate well into the soil. Typically, water sits on top of the soil in pools and accumulates in swamps or marshes where it is eventually lost to evaporation or drainage. As a result, highland vertisols are often only suitable for cereal cultivation. In these areas, crops are cultivated with the use of ox-drawn ploughs. Most of the land is used for grazing or remains fallow in areas where drainage is very poor.[14]

Vertisols become cohesive and present a problem for both mechanised and hand farming when wet. When wet and expanded, the soil is poorly drained and prone to cracking and contracting, which causes root damage when dry. However, despite the damage it does to root systems, the cracking of the soil does allow some water to trickle in and be accessed by plants. As the soil becomes wet, it swells, and the cracks close up. Larger cracks remain open longer, allowing more moisture to penetrate the soil. The soil usually becomes wet from the bottom-up, depending on how much water penetrates.[15]

Sugar producers at Wonji and Metehara in the Upper Awash have utilised this characteristic of vertisols, where vertisols are common. This natural cracking effect is more important in vertisols than accumulations of minerals in the soil. A build-up of exchangeable sodium affects vertisols differently than other soils. The amount of sodium saturation can be very high before it has an effect on agriculture. As a result, poor irrigation practices and the soil's low permeability have caused the levels of salts and sodium to build up to the point that the land can no longer be used for agriculture, as has happened in some areas of the Awash Valley. Other problems include many chemical reactions under the prevailing conditions of salinity and sodicity, a high pH, and deficiencies in nutrients, such as iron, manganese, zinc, and copper. If proper water management techniques are implemented sooner rather than later, many of these problems can be alleviated before they present too much of an economic cost or cause the land to be unusable. In areas where soil textures are coarser and proper water drainage measures are taken, these soils could be very productive due to the flat topography of the land and their ability to retain nutrients and hold water. Even so, because of their clay mineralogy, vertisols are vulnerable to erosion, especially when not covered with natural vegetation.[16]

Fluvisols cover 8 per cent of Ethiopia and are found in flat to nearly flat locations near water-based environments. Most alluvial deposits have their origins in the highland plateaus and are comprised of young volcanic soils. The various origins of alluvial deposits – different topographies, mechanisms of deposition, and climates – all create a wide array of fluvent soils varying in chemical and physical characteristics. As a result, there is a range of possibilities for natural vegetation and land uses on these soils. Fluvisols are usually found on alluvial plains, while eutric fluvisols can be found throughout the country. The large alluvial fans coming down from the northern Hararghe plateau, the southern Rift Valley, and western Ilu Ababor account for most of the fluvent soil areas. Calcaric fluvisols occur in the meanders and bends of the Awash, Baro, Omo, and Wabi Shebelle Rivers. Fluvents also occur around the margins of Rift Valley lakes, Lake Tana, the deltas of Lake Abaya, and Lake Turkana. Fluvisols are quite fertile, unless there is inadequate soil moisture or if salinity and alkalinity are excessively high. Another characteristic that makes fluvisols attractive for agriculture is the fact that they occur in flat terrains and near a supply of water. The soil often retains moisture for at least part of the year and sometimes can be irrigated afterward. During the rainy season, fluvisols are typically flooded, but can support intensive cropping once the floodwaters recede. Farmers in such areas typically cultivate cereals and legumes on exposed fluvisols. Fluvisols represent some of the most agriculturally productive land in Ethiopia, especially in the Awash River Basin. A major problem of fluvisols is the requirement for flood control and improved drainage. The mismanagement of irrigation can lead to problems, such as increased salinity and alkalisation, as is evident in some areas of the Awash River Valley.[17] In general, the greatest stresses on fluvisols are declining fertility and poor soil management, two things sharply affected by unpredictable rainfall. Fluvents tend to be unproductive in dry areas, needing large amounts of water and fertiliser. In these dry areas, pastoralism and small-scale farming are the main economic activities though the farmers find their soils becoming either too alkaline or saline.[18]

Soils containing the addition, decomposition, and accumulation of large amounts of organic matter in the presence of calcium characterise *mollisols*. The total amount of organic matter accumulated, its distribution, and rate of decomposition all depend on the vegetation cover and environment. Mollisols are typically dark in colour resulting from a high exchange capacity, high calcium content, abundant mineral colloids, and a high level of montmorillonite, all of which are associated with the accumulation of organic matter. The grassland and cultivated areas of Arsi and Bale (OR) and in the Middle Awash (AF) have mollisols.[19]

Mollisols typically occur in areas where there is sufficient moisture to allow for a significant accumulation of organic matter. They can be found on varying terrains, from flat alluvial to undulating plains to mountains. In Ethiopia, mollisols are associated with humid 'temperate' forests and high grass areas. They have large accumulations of organic matter, a high-to-medium base status, and deep leaching of calcium carbonate. Other features include a wide range of leaching factors, stages of weathering, and horizon differentiation.[20]

This soil group is among the most fertile in Ethiopia. Mineral stresses or deficiencies are uncommon. Major crops grown on these soils include tef, wheat, barley, fava beans, fodder oats, legumes, lupin, rape, mustard, linseed, and noog. They are often associated with alfisols in the transition zone between grassland and forest and drainage catena. Usually mollisols occur where there is sufficient precipitation to percolate through the soil. This causes soluble constituents and weathered material to move through the soil profile. This leaching process is similar to eluviation by solution but involves removal of materials from the entire solum.[21]

Luvisols are soils with an argic horizon – a subsurface horizon with higher clay content than the overlying horizon. These soils usually have brown to dark brown surface horizon with brown to strong brown or red argic subsurface horizon. Luvisols are mainly associated with gypsisols and cambisols that have an argic subsurface horizon. They also share common characteristics with vertisols by having 'slicken sides in the argic horizon' but not having vertichorizon. For the most part, luvisols can be good for agriculture; they have high base saturation and minerals that can be weathered quickly. However, luvisols with a heavily textured B-horizon can have permeability problems that hamper water movement and root distribution.[22] Luvisols can be found in Alefa, Quara, Chiga, and Metema weredas in North Gonder; Farta, Dera, and Esite in South Gonder; Bahir Dar Zuria, Achefer, and Adet weredas in West Gojjam zone (AM); Chole and Golocha weredas in Arsi zone; Goba Koricha, Habro, and Chiro weredas in West Hararghe zone; Bedeno and Melka Balo in East Hararghe zone; West Shewa and southern Borana zones (OR); and in eastern Gamo Gofa zone (SN). Luvisols occupy about 5 per cent of the country's land surface.[23]

B-horizon accumulation of low activity clays and low base saturation characterise *acrisols*. The top surface can be coarse-textured. These soils form mainly on sloping landscapes and humid tropical environments. Extensive leaching, low levels of nutrients, high levels of aluminum, and susceptibility to erosion make these soils less than ideal for agriculture. However, their pH of less than 5.5 can make them useful to grow acid-tolerant crops. Metekel zone (BE), western West Wollega, western Bale zones, northern Borana zones (OR), Gamo Gofa and Dawuro zones (SN), and Zone 2 in eastern Gambela have acrisols. Acrisols are estimated to occupy 5 per cent of the land surface of the country.[24]

Lithosols, Regosols, Xerosols, Yermosols, and *Solonchaks* occur in arid and semi-arid environments where precipitation is severely restricted, and productive agriculture is virtually impossible without irrigation. Occupying a combined 40 per cent of the surface areas in the country, they are the most extensive groups of soil in Ethiopia.[25] They form on various parent materials and are prevalent throughout the Afar and Somali states. Because of the slow weathering in dry climates, these soils usually occur on flat terrain, often in association with inceptisols and entisols. They differ from soils in more humid regions in that the main soil-forming process is an accumulation rather than leaching. The lack of vegetation means that the level of organic matter is low – often less than 1 per cent. Calcium carbonate and calcium sulphate tend to be in high concentrations, to the point that soil reactions are alkaline even if the parent material is acidic. Compared to soils from humid areas, these soils tend to be shallower, coarser, and have a weaker structure. Poor management of the soils can lead to soil degradation (especially through wind erosion, salinisation, and sodification); sodium levels in the soil are a particular concern, as sodium is toxic to plants. Under careful management, some of these soils may support arable farming with irrigation and pastoral activity.[26] At present, outside of the Wabi Shebelle and the Awash River valleys, there is very little arable farming in areas covered by these soil groups.

Farming systems

Ethiopia is one of the world's major gene centres. The variety of different agro-ecosystems in the country has given rise to a wide range of biodiversity in crops. Different landraces and wild varieties represent many crop species, such as sorghum, barley, tef, chickpeas, and coffee. Many are uniquely adapted to the local conditions. Farmers in Ethiopia use their knowledge of the local ecology to create farming systems adapted to the local environment.[27] Genetic diversity in crops allows farmers to take advantage of the microenvironments' soils, precipitation regimes, temperatures, altitudes, slopes, and fertility. Diversity also offers protection against diseases, pests, drought, and other stresses. Many plant and animal species are used for food, fodder, fibre, and medicinal purposes. Biodiversity also sustains production systems, improves human diets, and supports the livelihoods of local communities. Maintaining genetic diversity is thus crucial to sustainable agriculture, particularly for resource-poor farmers using low inputs on marginal lands.[28]

Farmers are not only crucial in helping create genetic diversity but also in sustaining its resources. Farmers usually save seed stock in containers, such as clay pots or rock-hewn mortars, as is done in central Ethiopia.[29] In addition to household storage, farmers also sustain seed supply through networks. One such network is the seed exchange in local markets. Despite the range of genetic diversity, including many primitive and wild types, genetic erosion is occurring at a fast rate. Several factors are causing such losses: replacement of indigenous landraces by new cultivars with high yield potentials (but low genetic diversity), changes in agriculture and land use, farmers' preference for crops that yield greater returns, loss of habitats, destruction of ecosystems, and natural calamities, such as drought.[30]

About 35 per cent of Ethiopia's land surface is considered arable land.[31] Of the country's total area of 110.4 million hectares, about 16.3 million hectares (14.8 per cent) are dedicated to cultivation. Around 56.4 million hectares (51.1 per cent) are used for livestock grazing. Forest cover is estimated to be about 3.9 million hectares (3.6 per cent). Other woody vegetation covers about 8.8 million hectares (8 per cent). Unproductive land accounts for 24.8 million hectares (22.5 per cent).[32] There are 12 million households engaged in agricultural activities in Ethiopia.[33] Ethiopian farmers have smallholdings. As the rural population grows and the non-farm sectors remain rudimentary, fragmentation of land into smallholdings is inevitable. In most rural areas, farm sizes are dwindling as farmers continue to subdivide holdings between their offspring. Nearly a third of the Ethiopian farmers operate on holdings of less than one-half of a hectare, 56 per cent have less than one hectare, and 83 per cent farm less than 2 hectares (see Table 2.1).[34] On average, the Ethiopian smallholder operates 3.2 separate plots of land. While fragmentation might consume more time, it also has advantages in a diverse agroecology. First, it lessens the peril of unmitigated crop failure. Second, it could help maximise productivity as a result of cultivating crops in different soil and environmental settings.

Farming systems vary depending on climate and altitude. Ethiopian farming systems can be broadly categorised into five major groups: mixed cereal-livestock farming, enset-root farming, pastoral livelihood, shifting cultivation, and irrigation agriculture. The mixed cereal-livestock farming system prevails throughout most highland areas, but with varying degrees of cereal-livestock combinations induced by differences in altitude and precipitation. In most areas, the land is intensely cultivated, and large varieties of crops are produced in vast quantities. The crops are primarily cereal, pulses, oil seeds, vegetables, and fruits. Root and tuber crops are produced in negligible quantities. Livestock are a vital part of the

Table 2.1 Per cent of farm holders by size of holding (hectare) by states

States	<0.10	0.10–0.50	0.51–1.0	1.01–2.0	2.01–5.0	>5.0
Tigray	7	25	28	28	12	*
Afar	56	16	12	11	5	*
Amhara	9	17	23	32	18	*
Oromiya	5	21	23	29	20	2
Somali	29	17	21	20	12	1
Benshangul-Gumuz	6	19	24	29	21	1
SNNP	7	45	27	17	4	*
Gambela	17	55	15	10	3	*
Ethiopia	7	26	24	27	15	1
Average holding	0.05	0.30	0.73	1.43	2.85	9.46

Less than 0.1.

Source: Central Statistical Agency. 2009. *Report on Land Utilization*, Volume IV, Statistical Bulletin No. 446, AddisAbaba: CSA, pp. 13–51.

cereal farming system. Each household keeps a mixed herd of cattle, sheep, goats, donkeys, mules, and horses. However, cattle are particularly highly prized because of their multiple uses. In Ethiopia, animal power has been used for thousands of years. Oxen are especially useful in providing draught power. Each household strives to own at least two mature oxen to pull the traditional chisel plough, or *maresha*, needed to prepare the seedbed. Oxen perform most the threshing; in fact, most of them are needed to tread on crop sheaves to separate the grain from the stalk. The amount of land cultivated very much depends on how many oxen a farmer owns. The availability of more draught power ensures ploughing of more land, timely cultivation, and more income. The breeding of cattle is geared toward satisfying this requirement. Cows produce oxen and provide supplementary income through the sale of milk and butter. However, not all farmers have a pair of oxen. Oxen are distributed unevenly; only one-third of the farmers own two or more oxen, about a third own only one ox, and the remaining farmers own no oxen at all.[35] Sheep and goats are used more readily to provide meat or are sold for cash in time of need. Donkey and mules are pack animals. Horses transport people and are also used as draught animals in North Shewa, Wollo, Gojjam, and Gonder zones.[36] Overall, livestock ownership is highly valued for the numerous benefits it provides to farmers, including:

> investment capital, available for use in contingencies, relatively visible; individual wealth creation (including for women); recurrent income (milk, meat and other products); manure (which, if supported by on-farm fodder, recycles nutrients at lower cost than inorganic fertilizers); energy (traction, transport); and productive uses for farm residuals (crop residues, browse, weeds, boundary plants, uncultivated grassland).[37]

Forage is not cultivated for livestock feed. Instead, livestock are grazed on communal pasturelands. The types and quality of pastures vary from altitude to altitude. Afro-alpine grasslands abound in the upper *Dega* and *Wurch* agroecological zones. Here the grasses are short, densely tufted, slow growing, and cold resistant. Species composition varies, although many species tend to be dominant. In the higher and colder moorlands of the *Wurch*, up to 4,000 m, *Koeleria capensis* and several *Penta schistis* sp. are common. There are more *Sporobolus olivaceus*, *S. africanus*, *Hyparrhenia arrhenobasis*, and *Panicum* sp. at lower altitudes. Temperate grasslands can be found in the *Dega* zone above 2,300 metres. They are frequently located in depressions and subject to seasonal waterlogging.

These pastures are dominated by mid-grasses, such as *Andropogon*, *Festuca*, and *Pennisetum*. The temperate grasslands are also characterised by an abundance of legumes, such as perennial and annual *Trifolium* sp., or clover, and annual *Medigaco* sp., or *Lucerne*.[38]

Sub-tropical savanna grassland occurs in the *Woyna Dega* zone, which extends from 1,500 metres to 2,300 metres of altitude. This type of grassland extends southward and eastward into the semi-arid lowlands that surround the highlands.

These grasslands have been grazed and seasonally burned for centuries. The species that dominate here are well adapted to frequent browsing. In the higher, moister locations, *Hyparrhenia* and *Themeda* sp. are common, while *Cenchrus, Aristida*, and *Chrysopogon* sp. can be found in the more arid regions. Herbaceous legumes and other nutritious forbs are less common on these grasslands.[39]

Another source of grazing and browsing for livestock during the dry season is forested areas. However, this is only possible where such forests still exist, such as the Arsi and Bale highlands and the western highlands of Ilu Ababor and West Wollega. Low, creeping perennial grasses cover the forest floor, especially where the forest canopy has been opened by illegal tree felling or girdling. Typical forest grasses are *Pseudochinolaena polystachya* and *Panicum* sp. Cattle and sheep graze these forest grasses. In the understory are various species of forbs and shrubs that provide grazing for goats. There are also seasonal movements of livestock. Lowland herds trek up to the highlands during the dry season, while highland herds move down to the lowland pastures during the wet season.

Crop residues are also sources of feed in the cereal-farming system. Residues usually take the form of cereal straw. Of the various cereal straws, only tef is nutritionally equivalent to medium-quality grass hay. Other cereal straws are poor-quality animal fodder.[40] In general, livestock feed is increasingly becoming scarce in the cereal farming system. As arable land is diminishing (due mainly to population increase, land degradation, and soil erosion), croplands have expanded and communal pastures have shrunk. Firewood shortages have also forced households to use crop residues as a substitute for fuel. Removal of crop residues from cultivated field leads to low organic matter in the soil. The absence of organic matter in the soils means the lack of essential soil organisms, such as fungi, bacteria, termites, insects, earthworms, and numerous small animals. These organisms are important soil makers.[41] Leaving crop residues on land also protect the soil surface from wind and water erosion. There is an inverse relationship between soil erodibility and organic matter, with a one per cent increase in organic matter reducing erosion by 15 per cent.[42] Returning crop residues to the soil also enhances atmospheric carbon sequestration in soils.

Livestock and agroforestry experts are promoting the planting of leguminous forage trees, such as *sesbania and leucaena*. Such trees can provide a valuable source of nutritious fodder during seasons when other forage is scarce. However, this use has yet to be widely accepted by highland cereal farmers. Sometimes such trees are planted as fences, but in general they appear to be regarded as ornaments rather than sources of forage.[43]

High density and rapid growth of the human population have put a strain on the land resources in cereal farming systems. Shrinking individual holdings, encroaching of cultivation of land available for pasture, shortages of feed for livestock, decreasing numbers of livestock, lack of draught oxen, and the use of manure and crop residues for fuel rather than fertiliser are all major problems that indicate this system is unsustainable.[44]

Enset-root farming systems prevail in parts of the southern highlands and the east and west Rift Valley escarpments, such as Kembata, Hadiya, Gurage,

Gedeo, Wolayita, Sidama, Kaffa, Shaka, Dawuro, and Gamo Gofa, among others. Enset is a crop related to the banana and has been cultivated in much of south and southwest Ethiopia for more than two thousand years. Historically, agricultural activities were mainly based on human labour and the use of the hoe, but the animal-powered plough has replaced the hoe in most areas. The importance of enset as a staple food varies from place to place. In some areas, enset is nearly the only staple crop. In others, cereals and tubers are dominant staple crops. There are also places where cereals are the principal staple crops, and enset is a minor source of nutrition. Other crops of major economic value include sweet potato, yam, potato, banana, peppers, pineapple, coffee, and field crops like chickpeas, lentils, and taro. Cattle are also well incorporated into the enset-root farming system; they are vital for draught, milk, breeding, and manure.[45]

Manure and compost are frequently used to replace lost nutrients in the soil. Enset benefits the most from the application of animal manure. Farmers often intercrop cereals with pulses to increase soil fertility. Crop residues of various sorts are animal feed. After threshing cereal straw from tef, barley and wheat are stacked and fed to animals during the dry season along with pulse crop residues. Where enset is the primary crop, enset leaves form a key source of animal fodder. Enset roots are also good sources of protein and are fed exclusively to lactating cows.

Land and draught animal shortages are major constraints in the enset farming system. The shortage of animal power is directly related to land shortage and lack of animal feed. People have converted land under woodlots and communal grazing into cropland, and there is little land left for further use. Population densities and growth rates exacerbate the situation.[46] Population densities are particularly high in enset-producing areas. Densities are 231 people per square kilometre in the Gurage Zone, 368 in Hadiya Zone, 386 in Wolayita Zone, 483 in Sidama Zone, and 535 people per square kilometre in Kembata-Timbaro Zone.[47] The average land holding size in these zones is less than half a hectare. Hence, an improvement in agricultural production through an increase in the land area is impossible in these zones. Only productivity gains and higher yields can enhance agricultural production in the region.

Pastoralism is the main mode of production in the arid and semi-arid lowlands, which cover more than half of the country (a subject to which Chapter 3 will be devoted). The lowlands are home to more than one-third of the nation's livestock population, which provides a livelihood to more than 10 million people. Southern Oromiya and much of Somali and Afar states are predominantly pastoral domains; most of the land is under communal grazing. The Afar, Borana, Somali, Nuer, and small groups in the lower rift valley pursue primarily a pastoral livelihood. The rangelands in these areas are primarily short grassland and acacia woodlands and are shrinking as more and more land is converted into farmland by means of irrigation and dry farming. As pastoralists are pushed away from their traditional pastures, they are forced to induce overgrazing on the remaining land available for pasture. Herd size is also beginning to decline, and pastoralists are altering the makeup of the animals they are rearing from grazers to browsers.[48]

Recurrent droughts, inadequate feed and nutrition, water shortage, and livestock diseases constitute major threats to the sustainability of pastoral livelihoods in lowland Ethiopia.

Shifting cultivation is practised in the Gibe, Didessa, Gojeb, Akobo, and the Omo river valleys. Both temperatures and precipitation are relatively high in these areas. Since soils are leached and inherently infertile, farmers have to shift from plot to plot periodically, usually after 2 to 3 years of cultivating a piece of land. Agricultural production is mainly for subsistence. Maize, sorghum, and finger millet are cultivated intermixed with beans, peas, pumpkins, yams, and gourds.[49] The Gumuz, Berta, Anuak, Mekam, and Mursi people are known as shifting cultivators but they also keep some cattle.

Commercial irrigation agriculture is also a growing farming system. There were nearly 800 irrigation schemes throughout Ethiopia with a total area of 200,000 hectares in 2006.[50] The total irrigate land increased to 1.9 million hectares in 2013.[51] The proportion of land uder irrigation is less than the irrigation potential of the country, which is estimated at 3.5 million hectares[52] with a possible maximum of 5.3 million hectares, including 1.6 million hectares potentially irrigable by harvesting rainwater and groundwater.[53]

There are several types of irrigation practised in Ethiopia. Traditional irrigation, which has been practised in Ethiopia for centuries, accounts for 38 per cent. Small-scale traditional irrigation operations are growing in many parts of the country, especially in the Amhara and Tigray regional states; in an effort to intensify agricultural production as the population grows, arable land becomes scarce, and the frequency of drought increases. Traditional irrigation uses physical structures built with local materials to divert river water. These structures are usually destroyed by floods during the rainy season and, thus, must be rebuilt each year. In the Amhara state, irrigated land increased nearly 200 per cent between 2001 and 2006. There were over 6,000 small-scale irrigation schemes covering a total of over 76,000 hectares of land.[54] Traditional irrigation is also quite common in peri-urban areas, especially in Addis Ababa, Bahir Dar, and Debre Berhan, for the production of vegetables for the urban market.

Modern communal irrigation accounts for 20 per cent. This form of irrigation was promoted after the 1974 land reform and the 1983 drought encouraged its adoption. Government and non-governmental organisations developed these irrigation schemes. The beneficiary farmers usually work through users' associations to operate and maintain them.

Modern private irrigation, which accounts for 4 per cent, began in the 1950s and 1960s when large tracts of land in the Upper and Lower Awash River Valley were given to foreign investors to run large-scale commercial farms. The farms were made possible with the construction of the Koka Dam and reservoir, which 'regulated water flows' of the Awash River 'with benefits of flood control, hydropower and assured irrigation water supply.'[55] Large-scale irrigation of the time included the Metehara sugarcane plantation (10,146 ha), Upper Awash fruits and vegetables (6,107 ha), Wonji/Shewa Sugarcane plantation (4,094 ha), and the Fincha sugarcane plantation (7,185 ha).[56] This category disappeared when the

Derg regime nationalised these plantations in the mid-1970s. After nationalisation, private irrigation systems were replaced by public ownership. Public-owned farms were expanded to include many irrigation schemes, including the Amibara Irrigation Project in the Middle Awash, the Alwero Irrigation Project in Gambela, and Gode-West Irrigation near Gode town in the then Hararghe province.

The post-1991 government stopped building irrigation projects and expanding state-operated agricultural schemes. The private sector was not interested in investing in large-scale irrigation schemes either. It was only in the mid-2000s that both the government and the private sector showed interest in investing in irrigation schemes. Privately owned irrigation-based agricultural schemes are now growing in the Awash, Meki-Ziway, Weito valleys, and Gambela and Benshangul-Gumuz regional states. As part of the Sustainable Development and Poverty Reduction Program (SDPRP), the government's interest has increased and has resulted in its financing of several ongoing major irrigation projects around the country (Table 2.2).[57] Additionally, the last decade or so has seen significant government, NGO, and donor investments in irrigation schemes in the highland regions of the country. Over 13,000 medium- and small-sized irrigation schemes have been developed in the Amhara, Oromiya, and Tigray regional states.[58]

Table 2.2 Ongoing major irrigation projects (as of 2010)

Irrigation Projects	Basin	Hectare
Kesem and Tendeho	Awash	90,000
Arjo-Didesa	Blue Nile	14,000
Humara	Tekeze	43,000
Gumara	Blue Nile	14,000
Lake Tana	Blue Nile	62,457
Koga	Blue Nile	7,000
Lake Abaya	Rift Valley Lake	31,920
Ziway	Rift Valley Lake	15,500
Wabi Shebelle	Wabi Shebelle	52,920
Raya Valley	Danakil	18,000
Kobo-Girana	Danakil	17,000
Wolkyte irrigation	Tekeze	40,000
Total		405,797

Source: Ministry of Water Resources, Irrigation and Drainage Project Ethiopia, Addis Ababa: Ministry of Water Resources.

Irrigation allows farmers to grow two or three crops per year where they once grew only one. However, water for irrigation is poorly managed. Flood or furrow irrigation is the prevailing method of irrigation. This technique is the most inefficient form of irrigation. Much of the water is lost to evaporation. Distribution losses are also substantial. Such losses should not be allowed to continue, and farmers need to find creative ways to efficiently manage scarce water. The introduction of drip irrigation system is very critical in this regard. Unlike flood or furrow irrigation, drip irrigation system involves watering plant root zones via a network of porous plastic tubes a few centimetres below the soil surface. This system has several advantages over surface irrigation: it saves enormous amounts of water lost to evaporation and inefficient distribution system. Studies have shown 'water productivity gains ranging from 91 percent to 149 percent have been achieved using drip irrigation'.[59] Drip irrigation systems also reduce weed growth and, thus, save time spent in weeding. Drip irrigation reduces leaching of nutrients from the topsoil, which mostly happens with surface irrigation. In addition, salinisation is significantly reduced with drip irrigation. To install drip irrigation system requires considerable investment up front. But a major commercial farming enterprise, such as the Wonji and Metehara sugarcane plantation and processing firms and the growing irrigation-based agricultural enterprises owned by foreign investors, should be able to afford it. Drip irrigation is extremely vital for water-scarce areas, especially the dry regions of the country where evapotranspiration is very high.

Commercialisation of arable land

The last few years have seen an increasing demand for farmland in developing countries by international and domestic investors to produce food crops for export. Africa is the principal target of this on-going global large-scale land acquisition by foreign investors. About 70 per cent (or 40 million hectares) of the commercial land deals that took place since 2008 occurred in Africa because it is assumed it has the most underutilised or unused land and, compared with Asia and Latin America, is low-priced.[60] This heightened interest in agricultural land as a potential investment was mostly triggered by the worldwide food shortages in 2008 following an increase in oil prices. Ethiopia is one of the 20 or more African countries that are selling or leasing land for intensive agricultural production on a large scale. Between 2004 and 2009, Ethiopia leased 1.2 million hectares of land to foreign and domestic large-scale investors.[61] Ethiopia plans to lease more than 3.6 million hectares of land to foreign investors for the production of food crops and biofuels.[62] The government has transferred over 200,000 hectares of land in Gambela regional state (Table 2.3). Also, well over a million hectares of the state's land are set aside for investors. Investors here pay only $1.15 a year per hectare of land. Gambela is the target because of its low population density and availability of rich farmland and water.

The government has also leased over 80,000 hectares of land in Benshangul-Gumuz regional state. It has earmarked about a million hectares (nearly one-fifth

Table 2.3 Regional distribution of land for investment

Regional States	Foreign investments in active projects (ha)	Identified for future investment (ha)
Addis Ababa	171	0
Afar	10,000	409,678
Amhara	20,702	347,430
Benishangul Gumuz	83,931	691,984
Dire Dawa	0	0
Gambela	202,012	1,238,005
Oromiya	214,003	438,212
SNNPR	79,770	529,181
Somali	N/A	N/A
Tigray	300	0
Total	655,907	3,654,491

Sources: Tom Lavers. 2012. 'Land grab' as development strategy? The political economy of agricultural investment in Ethiopia, *The Journal of Peasant Studies*, 39, 1, (January): 105–132; AISD/ MoARD, Amhara BoEPLAU, CSA, EIA, MoARD (2011), Oromiya Investment Commission, SNNPR Investment Agency (2008), Tigray EPLAUA.

of the state) for investment.[63] Most of the agricultural investors in Ethiopia come from countries like Saudi Arabia, India, China, the Gulf States, and South Korea. For instance, Saudi Star, owned by Sheikh Mohammed Hussein Ali Al Amoudi, a Saudi oil billionaire with vast holdings in Ethiopian commercial farming, has been granted a 60-year lease to cultivate rice on a 10,121-hectare of land in Gambela. Saudi Star also plans to lease another 250,000 hectares to grow maize and sunflowers. An Indian firm, Katruturi Global Limited, has received a 50-year lease on a 101,214-hectare, with another 202,248 hectares promised for expansion with the rights to abstract water for irrigation from nearby rivers.[64] Other India-based enterprises include Ruchi Soya, which has a lease on a 24,292-hectare of land to cultivate soybeans for edible oil production. Verddanta Harvests has gained a fifty-year lease on 3,036 hectares of natural forest to grow tea. Sannati Agro Farm Enterprise has signed a lease for over 10,000 hectares of land to grow rice destined for export to the United States.[65] The irony is that Ethiopia is relinquishing its fertile land, while it is persistently food insecure. The World Food Program provided 230,000 tons of food aid (worth $116 million) to Ethiopia between 2007 and 2011, while the Saudi Arabian investors were growing wheat, barley, and rice on Ethiopian land for Saudi citizens.[66] These investments do not boost food security because what they produce is for export. They have limited backward and forward linkages with the domestic sector and generate very little income and employment because they are highly mechanised productions.[67]

Foreign investors are attracted mainly by the low cost of land, water, and labour in Ethiopia. With a population approaching 100 million and growing, it also provides a substantial market for potential investors. The government can acquire land with ease because land is state owned. Critics argue that continued leasing of fertile land by foreign investors will lead to the evictions of poor farmers and pastoralists or their relocation to less fertile and easily degradable lands. Crop production geared for export also competes with domestic food production, and this can exacerbate hunger and undermine the social and political stability of the country.[68] But the Ethiopian government argues that such investments would improve infrastructure, increase crop yields, earn the country carbon credits, provide much needed foreign earnings, and create employment opportunities. According to the government, 2.5 million people were employed by the sector.[69] The government also argues that smallholder agricultural production is unproductive and needs to be complemented by growing large-scale farming, which requires large-scale capital investment. It believes that no matter how much assistance it provides its smallholder farmers, the country will not be able to feed its fast-growing population and considers that mechanised agriculture is the way forward. Also, the government contends that land leased to investors is non-cultivated and non-forested and will not lead to eviction of farmers and herders.

However, one might argue that those non-cultivated or non-forested lands 'may be used for grazing livestock or deliberately left fallow to prevent nutrient depletion and erosion'[70] as 'part of the cycle of shifting cultivation traditionally practiced' by the local farmers[71] that essential wetlands and local communities have user rights to such land. Others also argue that large-scale agriculture requires inputs such as chemicals, pesticides, herbicides, and fertilisers, and intensively uses water. All this would lead to monocropping and environmental destruction, including leaching of toxic chemicals to ground and surface waters, both near and far, and health risks from chemical exposure.[72] Monocropping depletes necessary soil nutrients, causes long-term degradation of arable land, and leads to habitat loss and reduced food security. Irrigation-based monocropping also puts a burden on water resources; it creates water scarcity and increases women's workload. Foreign investors secure not only the title to the land but also water rights when they invest in the land. Foreign investors usually abandon the country and go elsewhere once the land is exhausted and water for irrigation starts to dwindle.

There has been a growing concern about large-scale land acquisitions in developing countries. For instance, the United Nations Special Rapporteur on the Right to Food provides policy recommendations about agricultural investments in developing countries to protect both human rights and ecological integrity. According to Spieldoch and Murphy (2013), these policy recommendations include:

> 1. All negotiations should be conducted in a fully transparent manner and with the participation of local communities; 2. Any shift in land use can only take place with a free, prior, and informed consent of local communities concerned;

3. The rights of local communities should be safeguarded at all times through protective legislation at the national level; 4. Local populations should benefit from revenue generated from investment, and contracts should prioritize their development needs; 5. Labor-intensive farming systems, which create more employment should be given priority; 6. Investment should respect the environment through sustainable production practices; 7. Investors should be obligated to respect labor rights and a fair wage for farmers to benefit local communities; 8. Investors should provide a certain percentage of crops to be sold through local markets as a means to increase the productivity and benefits to local producers; 9. Investors should undertake impact assessments of the potential risks and benefits based on local employment and incomes; access to productive resources by local communities; the arrival of new technologies; the environment; and access, availability, and adequacy of food; and 10. Indigenous communities should be granted specific forms of protection to their rights to land.[73]

If implemented, the above recommendations could constitute a positive step forward in terms of promoting more socially and ecologically responsible agricultural investment deals. In reality, however, there is very little evidence that these recommendations are being implemented in Ethiopia. At any rate, with a growing population and the perpetual need for food security, Ethiopia cannot afford to continue giving away the use of its agricultural resource base (land and fresh water) to foreign investors. What Ethiopia needs is a socially and environmentally responsible agricultural development strategy so that the country's 'natural assets can continue to provide the resources and environmental services' on which current and future generations will depend.[74]

The country is also increasingly converting its prime farmland and resources to non-traditional export crops. Recent years have seen the development of foreign-owned large-scale cut-flower operations in the central rift valley region, mainly around the towns of Bishoftu, Mojo, and Ziway. In 2008, there were about 77 flower farms in the country, half of which were foreign-owned. The government says it has 2,000 hectares of land available to new investors in the industry. Investors are interested in flower farming due mostly to cheap land, suitable climate, soil and water quality, and low wages. An average flower farm worker earns about 15 Ethiopian Birr per day or about US$0.75. The floriculture industry also benefits from Ethiopia's relative geographical proximity to the Middle East and Europe and the availability of daily flights to the Middle East, Europe, Asia, and North America. In 2008, flower companies in Ethiopia employed more than 80,000 workers, of which 70 per cent were women.[75]

Modern airfreighting has also enabled the country to produce a wide range of perishable products, such as vegetables and fruits, for European and Middle East markets. In the last decade or so, output of export crops has been rising due to land and resources being diverted to them. Non-traditional crops are increasingly competing with established export commodities like coffee, tea, textile, and livestock products. For example, foreign earnings from the floriculture industry grew from

$660,000 in 2001 to $12.6 million in 2005 to $127 million 2007[76] to $178 million in 2009,[77] and its share of export earning has grown from less than one-tenth of one per cent in 2001 to 5.4 per cent in 2007[78] to 8.1 per cent in 2009.[79] This booming industry is expected to become the top foreign exchange earner, overtaking coffee. The projection is that foreign earnings from floriculture export will fetch the country nearly one and half billion dollars a year within the next few years.[80]

The flower industry is not without environmental problems. First, since the industry uses excessive amounts of water, the likelihood that it will deplete well waters is high. What is more, foreign companies engaged in growing flowers and other intensive agricultural productions are not charged for water. Second, these farms use large amounts of fertilisers, herbicides, and pesticides, and the potential for the contamination of the surrounding surface and ground waters is also high. Local environmentalists advocate for 'the use of red ash, biological pest control, and organic fertilizers' to counter the impacts of the business on the environment.[81]

Biofuels

Biomass use for energy is not new to the Ethiopian people. Biomass has long been used and is still being used in the form of charcoal, fuelwood, and animal dung. What is new, however, is that Ethiopia, like many other African countries, is seriously pursuing the development of biofuels to reduce its dependency on expensive, imported petroleum products, and diesel supplies and to strengthen its economy.[82] The government considers that the status quo is untenable because the country's total earnings from export commodities in 2008, for instance, could not even cover the cost of petroleum import, which was $1.6 billion for that year.[83] Petroleum import is expected to grow as the country continues to register impressive economic growth. The government argues that the country has sufficient land available to expand biofuel production without reducing the amount of food production. The government says biofuel production is taking place primarily on land with relatively poor soil fertility and low levels of rainfall. According to the government, the country has 23.3 million hectares of such land, of which 17.2 million hectares are in Oromiya state.[84] This marginal land can provide the potential for the cultivation of jatropha and other plants used in biofuel production. As of 2010, there were 83 registered biofuel investors in Ethiopia, of which 16 were operational. Land allocated for biofuel production in Ethiopia was around 600,000 hectares, up from 300,000 hectares in 2007. It is projected to increase to 2 million hectares in a few years, accounting for a mere one-tenth of one per cent of the existing cropped area.[85]

From the government's point of view, social, economic, and environmental benefits will accrue from biofuel production. Biofuels, it says, create jobs at various stages of its production: planting, processing, and transporting stages. Biofuel production is also expected to help break the country's dependence on costly petroleum imports and help with the balance of payments. Biofuels are considered an environmentally friendly alternative to petroleum-based transport fuels. They could help reduce the use of firewood and deforestation. Biofuels could also act as carbon sinks, and the country could earn carbon credits.[86] At present, the government plans to extract biodiesel from jatropha, castor oil, sunflower, and palm oil and ethanol from sugarcane.[87]

Proponents of biofuel development also see health and gender benefits related to the production of biofuels. The vast majority of the rural Ethiopians lack access to electricity and modern fuels, and they rely on biomass (wood, crop residues, and cow dung) to meet their household energy needs. Burning biomass fuels in open fires in poorly ventilated houses is a major source of ill health in Ethiopia. The smoke and chemicals emitted during burning contribute to respiratory disease. Women and their children are often at a greater risk of ill health from such exposure. The heavy burden on women and children to collect and carry firewood, and injuries caused by burns are additional health problems.[88] The use of modern biofuels significantly reduces these burdens.

There are also strong arguments against the promotion of biofuel production. Land used for growing biofuels is at the centre of the debate on food security and environmental impacts of biofuel production.[89] Large-scale biofuel productions compete with food production, and this can exacerbate hunger and undermine the social and political stability of the country. Almost every piece of land converted to biofuel production will inevitably compete for 'land, water, nutrients, and other resources with either food production or conservation.'[90] Biofuel production could lead to deforestation, destruction of wetlands, and displacement of small-scale producers and rural communities.[91] Many biofuel crops also consume large amounts of water. In Ethiopia, for instance, sugarcane dominates ethanol production. In the Awash River Valley, sugarcane is being produced using enormous amounts of irrigation water. It is a water-stressed region due to increasing agricultural activities, rising population, and declining groundwater levels. Additionally, growing biofuel production could have adverse impacts 'on biodiversity due to habitat conversion and loss, agricultural intensification, invasive species, and pollution.'[92] Moreover, not all biofuel production in Ethiopia happens to use marginal lands. Kassaye's study shows that in Wolayita Zone (SN), 20,000 contract smallholders produce castor oil and jatropha for a foreign private company known as the Global Energy Pacific on prime agricultural land. The production is done at the expense of locally consumed crops, such as maize and sweet potato.[93] With over 400 persons per square kilometre, Wolaita Zone also happens to be one of the most densely populated zones in the country.[94] In addition, biofuel production is a serious concern in the context of feeble and contested local land and resource rights.

Challenges to sustainable agricultural development

Climatic vagaries

Ethiopia's size and geographical diversity mean that it has a wide array of topographic features and a range of climatic zones. Climatic zones are traditionally categorised based on temperature and moisture, ranging from the hot and arid lowlands to the cold and humid highlands. Ethiopian agriculture is mostly rain-fed. Rainfall variations cause considerable volatility in agriculture growth and productivity. Rainfall tends to be variable and unpredictable in terms of both time and space. Rainfall variability reduces yields and leads to land degradation. Yields typically fall whenever there is too little rain, but excessive rain and flooding can also damage yields. One of the greatest threats to the country's livelihoods, in particular, and ecosystems, in general, is climate change. Climate change is brought about locally through regional vegetation change due to poor resource management. Climate change is also brought about globally by the rising concentration of carbon dioxide and other greenhouse gases in the atmosphere, including methane, nitrous oxide, and hydro-fluorocarbons.

Ethiopia's climates are both varied and varying. They range from humid tropical regimes in the southwest, central, and eastern highlands to semi-arid regimes in the southern and eastern regions to arid and hyper-arid regimes in the northeast and extreme southeastern regions of the country. All of these climatic regimes display differing levels of temporal variability in precipitation. Over the last half century or so, this climatic variability has manifested in extreme weather conditions, such as droughts and floods, which have increased in frequency and severity.

From a meteorological point of view, three seasons are identified in the country. *Belg* is a short rainy season that occurs between February and May as a result of a southeast airflow from the Indian Ocean. The rains are essential for short-cycle crops. These rains are also essential for preparing the land for the *Mehr* long-cycle crops and replenishing groundwater in lowland pastoral areas. *Kiremt* is the primary rainy season in most parts of the country. It lasts from June to September, during which the Atlantic moisture system brings heavy monsoonal rainfall to southwestern, central, and northern Ethiopia. The *Bega* is essentially a warm, hot, dry season that lasts from October to the end of January. Only occasional brief rains occur during the *bega* season, particularly in the south and southeast. The *bega* is a harvest season for *Mehr* crops.[95]

The amount and timing of precipitation vary from place to place depending on latitude, altitude, topographic orientation, and location in relation to incoming moisture-bearing winds, distance from oceans or other major water bodies, and other local conditions. Rainfall decreases toward the north, with the south and southwest regions being the wettest. The areas with the highest precipitation occur in the Illu Ababor, Jimma, and Kaffa zones in the southwest. This region has a mean annual precipitation, which is about 2,200 mm though it can exceed 2,800 mm in other areas. For the most part, precipitation occurs year-round. This region is not the highest in Ethiopia but receives the most rainfall due to its location rather than its altitude. During the high-sun season in the northern hemisphere, moisture-laden air from the Atlantic blows over the Congo Basin into southwestern Ethiopia. The air masses gradually lose moisture as they move northward, and less rain falls. Precipitation ranges between 1,100 mm and 1,600 mm per year on the highlands of East Wollega, North Shewa, East Gojjam, North and South Gonder, and North and South Wollo zones. Further north, the Tigrayan highlands receive less yearly rainfall, ranging between 500 mm and 900 mm. On the plateaus of Axum, Adigrat, and Mekelle, the average annual precipitation is about 980 mm. The eastern half of Tigray is on the leeward side of the source of moisture and averages between 400 mm and 600 mm. The southeastern highlands receive an average annual precipitation of about 800 mm to 1,200 mm. There is a remarkable concentration of rainfall in all of the highland regions during the months of June, July, August, and September. Precipitation over the four-month period amounts to more than three-quarters of the annual total. Overall, the Ethiopian highlands receive adequate precipitation. Exceptions are some of the overexploited highlands in the north, the leeward sides of mountains, and the northeastern Rift Valley slopes.

Lowlands receive a much smaller and less frequent amount of precipitation than the highlands. The amount of precipitation and seasonal concentration also varies. The southeastern lowland areas, covering nearly the entire Somali Regional State, average fewer than 400 mm of rainfall, with the wet seasons in autumn and spring. The eastern slopes of the Harar-Bale highlands, overlooking the southeastern lowlands, and eastern Borana receive an average of 400 mm and 800 mm. The aridest region of the country consists of the low-lying areas of the northern Awash River Valley in the Afar Regional State. Annual rainfall in these regions averages less than 200 mm, hardly enough to pursue farming activities.[96] The scant rainfall in this region is erratic and vulnerable to high year-to-year fluctuations. The land is prone to high rates of evapotranspiration due to high temperatures as well, which reduces the availability of surface water for agriculture and domestic uses.

Rainfall in Ethiopia is not only seasonal but also variable, both during seasons and between years, due largely to topographic effects and weather anomalies. Rainfall is unpredictable in the amount, time, and duration from year to year. Sometimes rains start later than usual, other times they come too early. Sometimes the rains are concentrated over a short duration, and other times they end too soon or are interrupted by long dry spells in July or August. Overall, the deviation from the mean annual rainfall is 14 per cent in the wettest region of the country (the southwest), 30 per cent in eastern Amhara and Tigray states, and 50 per cent or more in the northeastern Afar and Somali Regions. In general, the larger the total annual rainfall a region has, the smaller the year-to-year variability. Rainfall variability is extremely important for agricultural purposes as crops primarily fail because of early, late, or insufficient rains.[97] Additionally, the amount of water available for crops is not always the amount of rain that has fallen at any given time since some water is lost to evaporation and runoff. High rates of runoff and evaporation lead to a decrease in soil moisture and an increase in soil erosion and salinity.

Extreme climate events are common in parts of Ethiopia. For centuries, the country has been subjected to excessive climate variability and severe weather, especially true of the northern highland areas. Large-scale droughts that resulted in the deaths of vast numbers of people and livestock and massive displacement litter the history of this region. Such events have occurred on average three to five times during any given century, but weather patterns have grown more severe over the last half century. Droughts occur regularly.

Dry seasons are becoming ostentatiously longer. Wet seasons are also becoming shorter. Even areas that normally enjoy copious rainfall in the past are now experiencing a sporadic drought. Climate variability exists even within the same region. This variability explains why there are many drought-stricken pocket areas in otherwise good crop years and why some highland areas are not affected by harsh climatic events. Some areas that used to be drought-prone have experienced 'high and well-distributed levels of rainfall' in recent years.[98] Flooding occurs frequently in both wet and typically dry areas. Many parts of the country are experiencing scorching temperatures and intense sun. Ice no longer covers high mountains like the Ras Dashen. Many climate experts believe these events are probably the result of warming temperatures brought on by anthropogenic global climate change.

Over the last half century or so, the Ethiopian climate has turned eerie and unpredictable. Ethiopia is one of a hundred countries in the world that are most vulnerable to the effect of climate change. The UNDP Climate Change Profile for Ethiopia showed:

> the mean annual temperature increased by 1.3 degrees Centigrade between 1960 and 2006. The temperature increase has been most rapid from July to September (0.32 degree Centigrade per decade). The average number of hot days per year has increased by 73 (an additional 20 percent of days), and the number of hot nights increased by 137 (an additional 37.5 percent of nights) between 1960 and 2006. The rate of increase is seen most strongly in June, July, and August. Over the same period, the average number of cold days and nights decreased by 21 (5.8 percent of days) and 41 (11.2 per cent of nights), respectively. These reductions have mainly occurred from September to November.[99]

However, long-term rainfall trends are hard to detect because of the 'high inter-annual and inter-decadal rainfall variability'.[100] The years from 1960 to 2006 saw very little changes in any season in terms of the mean annual rainfall. Rainfall decreased from July to September in the 1980s but recovered in the 1990s and 2000s. The lack of sufficient daily rainfall records makes it difficult to detect 'trends in daily rainfall variability and changes in rainfall intensity'.[101] Data from the National Meteorological Agency of Ethiopia also indicate that the country had 'experienced 10 wet years and 11 dry years over the last 55 years,' demonstrating the strong inter-annual variability.[102] The Famine Early Warning System Network's rainfall analysis shows that rainfall during the March-May rainy season has been declining in southern and eastern regions of the country since 1996. This decline in rainfall has been linked to recent warming in the western Indian Ocean.[103] If the severity of the climate continues to increase, the country could suffer from diminished agricultural production, water shortages, an economic downturn, population displacement, and loss of lives. According to experts, Ethiopia is estimated to lose cereal production potential in the 2080s owing to adverse climate change.[104] The 1973–74 famine took the lives of over 200,000 people and reduced many more to destituteness. The 1983–85 famine was perhaps the worst in Ethiopian history, resulting in the death of no less than two million people and countless animals. The 1998 El Nino flood contributed to a huge loss of livestock, crops, and income. The 2015–16 drought, due, in large part, to the El Nino global weather pattern, affected no less than 15 million people in the country. In the age of climate change, Ethiopia is likely to see many more such episodes. Future disasters induced by climate change could be worse.

Flooding has also been a frequent problem in many parts of the country in recent decades, largely due to deforestation of catchment areas and poor land use. Floods have become a recurring problem in many parts of the country. Destructive flooding caused considerable loss of life and property damages in 1988, 1993–96, and 2006. In Dire Dawa, in August 2006, heavy rain in the surrounding highlands led to the worst flooding in several decades and brought devastation to communities living on the banks of the dry riverbed that cross the city. It displaced upward of 3,000 people, with more than 300 people killed.[105] During the same year, flooding in West Shewa and South Omo also killed over 600 people and displaced more than 35,000 people.[106] Water-borne diseases associated with floods claimed many more lives. Flooding episodes are also not infrequent in Western Tigray, the Afar and Gambela states, and the Lake Tana catchment areas.[107]

Many drought events in the country may not have been solely caused by deforestation and land-use change as many experts have argued but by changes in global ocean temperatures, as Alessandra Giannini, a research scientist at the International Research Institute for Climate and the Society and the National Science Foundation's most prestigious career award winner, has convincingly concluded in her recent research.[108] Giannini and other scientists attribute the meteorological causes of drought to 'the quasi-oscillation of atmospheric circulation triggered by sea surface temperature (SST), anomaly over the southern Pacific, and associated El NINO/Southern Oscillation (ENSO) events, together with SST anomalies over the southern Atlantic and Indian Oceans'. ENSO events weaken and dislocate rainfall in Ethiopia and the Horn of Africa, causing droughts.[109]

Even though early warning systems for climate, water, and weather-related hazards have managed to save many lives and reduce damages to properties in developed economies, such vital technologies are yet to be readily available to developing countries like Ethiopia. In many developed and developing nations, technologies predicting climate changes have been used in many useful applications, including rainfall predictions to guide agriculture, water resources and fishery development, health, and natural hazards applications.[110] Ethiopia is already facing climate variability and change and desperately needs access to such advanced technologies. Early warning systems and related climate change predicting technologies could provide information about weather and climate fluctuations and extreme events, such as drought and flooding. Such technologies allow the country to develop effective adaptation measures to prevent the most horrible consequences of future climate change. The government's Disaster Prevention and Preparedness Authority keeps track of the food security situation in the country and forecasts looming drought so that it can take proper measures. Unfortunately, the authority lacks adequate resources to invest in early warning advanced technologies to accurately and promptly warn of disasters before they occur. It will take years and a huge government investment in agriculture and international assistance to access advanced technologies before rural Ethiopians experience an existence less vulnerable to climatic vagaries.

Land degradation

Soil erosion is one of the most destructive forms of land degradation. Soil erosion removes not only soil particles but it can also 'reduce the abundance and diversity of soil biota by physically removing organisms, destroying their preferred micro-habitats and changing micro-climatic conditions within the soil.'[111] In addition, it can reduce the amount of soil water available for plant growth. Erosion can take place without human intervention, but human activities have accelerated it by pursuing inappropriate land use and by removing the protective natural cover. Soil erosion occurs mainly in the rainy season in the form of water erosion and in the dry season in the form of wind erosion. Much soil is lost due to erosion in Ethiopia, as is evident by rills, desiccated farms, gullies, and rivers brown with sediment. Erosion has severely degraded more than half of the country's arable land.[112] The estimated national average soil loss is 12 tons per hectare per year. The estimated soil loss on pastureland is five tons per hectare per year, 42 tons per hectare for cultivated land, and 400 tons per hectare for land without vegetation cover.[113] The total estimated loss of soil from highland Ethiopia is 1.5 billion tons annually, accounting for a loss of 1.5 million tons of grain each year.[114] Heavy rains in the highlands during the June–August season also cause rivers in lowland areas to overflow and erode extensive amounts of precious topsoil every year. International rivers such as the Blue Nile and Wabi Shebelle carry away much of the eroded soil. The sediment load carried away by the Blue Nile River alone is estimated at 140 million metric tons annually.[115] Most soil erosion occurs on cultivated land, usually in the form of sheet and rill erosion. Erosion also occurs in grasslands and forests because of overgrazing and deforestation.[116] According to the Global Assessment of Soil Degradation (GLASOD), 26 per cent of Ethiopia's land suffers from degradation, affecting about 29 per cent of its population.[117]

The causes of soil erosion are multifaceted and have natural and socio-economic dimensions. Rainfall is the principal factor in sheet and rill erosion, and the amount of rainfall a place receives affects erosion rates. The intensity of a rainstorm is also related to the amount of erosion. Annual precipitation in Ethiopia ranges from almost nothing in the Danakil Depression (AF) to more than 2,400 mm in southwestern Ethiopia (which includes parts of western Oromiya, SNNP, and Gambela states). Highland areas in these and other regions of the country receive most of the seasonal heavy and intense rains. As a result, water erosion is the primary cause of soil erosion in the highlands because of the heavy concentration of rainfall over three to four summer months. Over 72 per cent of the highlands receive 600 or more mm of rain from May to September.[118] The hard-beating, intensive summer rains in the highlands cause severe erosion, especially when land is without vegetation cover. Slope gradient and slope length are also important components of soil erosion in highland regions. Cultivation of steep slopes and watershed areas is quite common in highland Ethiopia. The steeper the slope, the higher is the potential for erosion. In one case study, in North Shewa Zone (AM), losses were 130 tons per hectare per year on 0–15 per cent slope, 152 tons per hectare per year on a 15–30 per cent slope, and 170 tons per hectare per year on a 30 or higher per cent slope.[119] About 21 per cent of the highlands have slopes less than 8 per cent, 60 per cent have slopes over 17 per cent, and 33 per cent have slopes greater than 30 per cent.[120]

Land use is another major factor in sheet and rill erosion, with the primary problem being the disappearing canopy and surface cover. Crop types in cultivated areas and the tilling of the soil also play a part in the level of erosion, as different crop types will provide more or less canopy for the soil. Fine or coarse tilling will also affect the ability of the soil to be eroded. For example, perennial crops, such as coffee and enset, cause less erosion than annual crops. Perennial farms help to shade the soil and maintain the moisture level.[121]

The time span of human occupation on the land is another factor. Those regions with the longest human habitation experienced the most soil degradation in the country. The highlands in the Amhara and Tigray states are the oldest inhabited areas. Most of the forests there were cleared long ago to make way for farmland, to provide building materials, and to provide firewood. As arable land dwindled, more and more people had to push onto steeper and steeper slopes. Once the natural vegetation is removed, the soil is easily eroded, and the land is exposed to heavy seasonal rains. The land shortage also meant that arable land was cultivated continuously without rest – resulting in loss of nutrients, water-holding capacity, and diminishing yield.[122] Losses of soil nutrients from farmland mean more erosion and less productivity. The level of nutrient losses by erosion is very high in highland Ethiopia, primarily due to energy mining of animal manure and crop residues through burning. According to Ethiopian soil scientist, Demel Teketay, there is an inverse relationship between erodibility and organic matter. The increase of erosion by 15 per cent causes a loss of 1 percent organic matter.[123] The use of animal manure and crop residues for fuel, instead of soil cover and fertility, has reduced crop yields in much of highland Ethiopia. Land productivity has also declined with the shortening of fallow and crop rotation periods and as people rely on marginal lands more heavily.[124]

Over the centuries, smallholder and independent farmers have developed various methods of traditional soil and soil moisture conservation practices to refurbish and maintain soil fertility for sustainable food production. These methods were developed through experiments, observations, and practices under diverse ecological conditions. In highland regions of the country, for instance, stone bunds or stone terraces were used to capture the sediments washed downhill during the rainy season. Farmers built a single-line or double-line stone bund depending on the degree of slope and availability of stone. The single-line bund was used when the slope has a gradient of 20 per cent or more and stones were scarce, while the double-line bund was built when the slope has a gradient of less than 20 per cent and stones were more plentiful. Farmers also left a series of unploughed strips, up to one-half metre in width, in fields ploughed on slope land. The strips captured eroded soil, reduced flood risks, and grew animal feed. To prevent topsoil erosion from run-off and flow and to provide drainage for waterlogged crops during the rainy season, a set of parallel ditches were dug in ploughed fields. The ditches were spaced based on gradient, with steep cultivated land having many more ditches than a shallow land of equivalent cultivation.

Other conservation methods, such as contour ploughing, intercropping, and fallowing, have also been practised for centuries in many Ethiopian farming communities.[125] Contour farming, 'plowing and planting across the changing slope of the land, rather than in straight lines,' reduces soil erosion and helps retain water in the soil.[126] The most important traditional conservation structures are the contour stone terraces, which are used in Konso. These measures are part of a highly productive agricultural system that provided a diversity of crops while ensuring continuous protection against soil erosion.[127] Intercropping four to five different crops on the same plot of land is common in the humid areas of southwest Ethiopia. For instance, Konso and Wolayita zones (SN) intercrop carbohydrate-rich maize that depletes soil nitrogen and protein-rich haricot beans and fava that add nitrogen to the soil. People grow cereal with legume crops in the rotation.[128] Crop rotation provides many benefits: it can help maintain soil fertility and conserve soil as well as control diseases, pests, and weeds. In addition to adding organic matter, closely grown grains, such as tef, wheat, barley, and oats, can provide vegetative cover to reduce soil erosion and runoff. Crop rotation also helps to sever insect and disease cycles. Leguminous crops, such as beans and chickpeas, augment nitrogen levels in the soil,[129] thus reducing nitrogen and other fossil-fuel-intensive inputs. Fallowing helps soil restore nutrients and recover its fertility, reduces erosion, controls weed growth, and limits the spread of plant pests and diseases.[130] However, fallowing is no longer practised in densely populated rural communities because of land scarcity.

Agroforestry – the practice of growing trees among crops – also has long been practised in Ethiopia. It is especially true in areas where population density and land scarcity have caused intensive land use. Agroforestry practices include cultivating specific tree species, soil nutrient building, providing forage, fuelwood, and medicinal herbs.[131] Farmers have long practised an integrated production system, involving a mixture of woody perennials, crops, and domestic animals on the same land unit. Planting woody perennials on farmland has many advantages. The trees can protect crops from the elements, prevent wind and water erosion and run-off, improve the nutrient level in the soil, maintain soil composition, and provide fruit, animal feed, wildlife habitat, and wood for burning and building,[132] and enhance carbon sequestration in the soil. Harrison points out the benefits of such a system in central Ethiopia:

> In the plain around the town of Debre Zeit [...] the fields are scattered with *Acacia albida* trees. They are not planted but fostered by farmers whenever they sprout up. The tree keeps its leaves in the dry season and serves as browse for livestock when little other fodder is available. At the end of the dry season, the leanest time of the year, it drops protein-rich pods – up to 5 tons per hectare – that can be used for fodder. When the rainy season arrives, the tree obligingly sheds its leaves, so it does not compete with crops for light. The trees' nitrogen-fixing root nodules, leaf fall, droppings of livestock that seek shelter from the hot sun enriches the soil around the trees. Crop yields around the trees "are double those that are farther away."[133]

As observed in the Hararghe Highlands, grain yields increase when farmers plant cereal crops under *Acacia albida* compared to out in the open. Farmers in Shashamene, around Hawassa, Wolaita, and Gamo Gofa also grow *Acacia albida* with cereals. Associated with *Acacia albida*/cereal crop systems are silvopastoral systems. In these, cattle, sheep, and goats are grazed under *Acacia albida* and *Acacia abyssinica*. Other species used are *Moringa oleifera, Croton macrostachys*, and *Ficus* spp. *Moringa oleifera* is grown on terraced farms and is prized for its edible young leaves, fruit, and medicinal values.[134]

Other forms of traditional agroforestry include multistoried and multi-purpose tree systems in which hops, beans, banana, false banana, coffee, khat, *Croton macrostrachys*, *Cordiaafricana*, *Vernonia amygdalina*, *Albizia gumifera*, *Ziziphus* spp., *Juniperus procera*, *Podocarpus gracilior*, and *Erythrina abyssina* are all grown together. There are also some eucalyptus and other plantations that are grown in undefined spatial arrangements. Cactus, euphorbia, and eucalyptus are also sometimes used for shade and fences. Such fences that demarcate individual holdings and small woodlots are common in Kembata, Wolayita, Gurage, Sidama, Keffa, and Konso (SN). However, these species are more prevalent on cultivated land because there is little control on communal lands, and people often cut trees for fuelwood or timber.[135]

Even though Ethiopian agriculture has shown some improvement in productivity in recent years, it is one of the slowest increases in the world. Low soil fertility contributes to low productivity in agriculture. By and large, most Ethiopian soils are low in nitrogen and deficient in phosphorus, potassium, sulfur, and magnesium. Soil fertility-enhancing inputs can do much to improve yields, but few of these inputs are available for most Ethiopian smallholders, despite efforts from the government and donors to provide chemical fertilisers and pesticides. The country cannot afford (both financially and ecologically) to continue depending on conventional high-input systems based on commercial chemicals. Farmers should be encouraged and supported to adopt low-input farming by using resources found on or near the farm, such as compost, biological fixed nitrogen, biological pest controls, and other nutrients released from organic matter and green manure. Yield might be lower than a high-input farming system, but the lower cost can make it profitable. Such a system is not only equally if not more profitable, but also causes less erosion and pollution, enriches the soil, is more energy efficient, and relies less on state subsidies than the conventional high-input system, and is more sustainable.

Diminishing holding size

Farmland is coming under greater pressure from growing rural population and recent expansions of biofuel and large-scale commercial production. Actual farm holdings continue to diminish in size. Declining plot size will make it increasingly difficult for households to subsist on agricultural production. About three-fifths of the Ethiopian farmers cultivate less than 2 hectares of land, as compared to 24 per cent in Uganda, 13 per cent in Namibia, 11 per cent in Burkina Faso, and 7 per cent in Senegal.[136] As the land passes over to the new generation of farmers, it results in further fragmentation. Since a land allocation is restricted to the place of residence, areas of high population density are often divided over and over until the pieces allocated are barely enough for subsistence. This situation, in turn, leads to 'diseconomies of time, labor utilisation, and scale.'[137] As the availability of arable land per capita declines, it becomes imperative that farmers produce a higher output per unit area of land than before. In addition, expanding rural employment opportunities in non-agricultural activities (for instance, by creating environment-based community projects, such as soil conservation, afforestation, micro-dams, water supply, road construction, and irrigation management) can reduce the pressure on the diminishing holdings.

Land tenure security issues

In Ethiopia, ownership of land is in the hands of the state. Farmers exercise only usufruct rights. In other words, they have the right to use but cannot sell or mortgage the land. Farmers make their decision about what to grow on the land. The farmer can use the land as long as he/she lives. His or her heirs inherit use right when he/she dies. The government views public ownership of land as increasing social equity and decreasing landlessness. The government fears that the privatisation of rural land will lead to the return of a class of landless poor, as was the case prior to the 1974 revolution. The concern is that distress sales may occur with privatisation and dislocate farmers from their land. In agrarian countries, land is likely to hold its value against inflation. It may prompt individuals to buy large amounts of land as investments against inflation rather than for agricultural purposes.[138] At any rate, the government insists that under public ownership, the rights of individual smallholders are protected to ensure that they are not arbitrarily overridden by the state. According to the government, land can only be expropriated with due process of the law and under exceptional circumstances.

There are many strong views about the prevailing land ownership and tenure system. Critics of the government's land policy, however, argue that tenure insecurity will adversely affect an investment in land.[139] In their opinion, privatisation of land would allow farmers to invest in land, use land as collateral to access credit, and become mobile if they so desire. The current policy also ties the rights to land to the owner's place of residence. As soon as the landowner leaves the area, he/she loses the rights to the land, inhibiting the 'free movement of people and the development of land markets,' according to Dessalegn Rahmato.[140] The government seems to recognise the likely undesirable effects of tenure insecurity. To that end, it has passed laws allowing regional states to issue a certificate – not title – of sustained 'ownership' of the land to farmers. In 2009/2010, 12.6 million rural household heads had land right certificates.[141] In 2014, 99 per cent of Tigray, 95 per cent of Amhara, 92.6 per cent of Oromiya, and 76.8 per cent of the SNNP household heads had land right certificates.[142] Critics, however, argue that since land right certification does not allow the transfer of land, it does not ease the restriction on a free land market. It is, therefore, questionable whether this policy would make farmers secure enough to make long-term investments in land or practise sound land and resource management. They also point out that smallholders' concern regarding future land redistribution can cause tenure insecurity.

Whether the current tenure land system continues or land is privatised must be viewed in the context of broader changes in the rural sector. Mersha and wa Githinji argue that, rather than expect development to be driven by a surplus in the rural sector, the government must first transform the agricultural sector, boost its productivity, and offer alternative forms of employment. They advocate an approach that focuses on the sequencing of agrarian reform. For land reform to be successful, they argue, it must be accompanied by a series of other reforms. If the intention is to reduce poverty, these other reforms must take place first. They suggest that the government should do four things to accomplish this goal: first, invest in labour-intensive public projects aimed at improving rural infrastructure and increasing agricultural productivity. Access to markets and infrastructure is crucial to reducing poverty. Such schemes would include, but not be limited to, roads, irrigation structures, rural electrification, terraces, and reforestation. As these projects lower the cost of farming, individuals may also begin to make private improvements on the land. Such projects would not only increase agricultural productivity but also provide an alternative source of employment, which would ease population pressure. Small industries in rural areas would also cost less and would have the effect of making future long-term non-agricultural employment sustainable. Second, the government should implement a national campaign to improve literacy and education. Like improving infrastructure, a campaign to improve literacy and education offers both employment and increased productivity. Third, the government should seek to intensify and extend agricultural inputs and extension services. The administration needs to increase poor farmers' access to draught animal power. Fourth, introduce new commercial crops, where possible. It is also critical to note that the forward and backward linkages created by industrialisation should be a factor in selecting which crops to introduce.[143]

Agricultural pests and vermin

Ethiopian farmers cope not only with weather-related factors, such as rain, drought, and floods, but also with a host of biological factors, including disease and insect infestation and other environmental threats and conditions surrounding their livelihoods. Agricultural experts estimate about 40 per cent of the yield of eight of the most important food crops worldwide are lost to insect pests and diseases.[144] In Ethiopia, pests and vermin exact a serious toll on crops, both as the crops grow in the field and after harvest. No one knows the exact magnitude of crops lost to pests, but various animals, birds, insects, and diseases attack almost all of the crops produced in the country. Farmers face substantial losses, both before harvest time in their fields and after harvesting in granaries and other storage facilities. The estimate for pre- and post-harvest crop losses is anywhere between 11 per cent and 25 per cent.[145] Weeds also affect crop yield. Weeds use up significant amounts of soil moisture and nutrients. Some weeds are very problematic and take great effort to eliminate – or even reduce their impact. Many farmers are unable to weed their fields at the proper time due to labour shortage.

The most common and serious insect pests in the highland regions include African armyworms, grasshoppers, Wollo bush crickets, stalk borers, bollworms, cutworms, termites, shoot-flies, red tef worms, snout beetles, plusia worms, aphids, sorghum chafers, and thrips. Birds such as the red-billed quelea, black-headed weaver, chestnut weaver, northern masked weaver, and bishop's sparrows are major pests. Rodents, including *Arvicanthus abyssinicus*, *Arvicanthus dembensis, Rattusrattus, Thryonomys* spp., *Praomys natalensis*, and *Tachyoryctes splendes*, also cause substantial crop losses. Storage pests, such as maize weevils, red flour beetles, pea weevils, and bruchids, also incur substantial losses to farmers' output. Fusarium, smut, rust, blight, anthracnose, and leaf blotch are among the most destructive plant diseases. Plants that are valueless take up precious nutrients and moisture. Between 1995 and 1997, the Amhara regional state alone lost 1.6 million quintals of grains to various agricultural pests, including insects (39 percent), plant diseases (12 per cent), weeds (26 per cent), birds (1.6 per cent), rodents (1.4 per cent), and storage pests (20 per cent).[146] Farmers bear the full burden of all these menaces because they have no access to agricultural insurance and receive little or no help from the government to recuperate.

Conclusion

The economic and livelihood foundation of Ethiopia depends largely on primary resource extraction and export activities in agriculture. However, the country's agricultural resources are under increasing pressure from growing rural population, growing urban demands, and increasing conversion of land for the production of biofuels and large-scale commercial food production for export. Land has been subdivided to accommodate more family members, thus reducing the plot to impracticable size to sustain farm households for their livelihood needs.

Agricultural productivity is extremely low due largely to the lack of adequate access to essential inputs, limited new agricultural technologies and innovations in farm practices, and poor infrastructure. There are environmental factors that all Ethiopian farmers face, such as drought and floods. Farmers lose a substantial amount of their crop yields to insects, weeds, pathogens, mammals, and birds. Unsustainable land use in rural areas is causing land degrading, including soil erosion, nutrient depletion, salinity, and loss of biodiversity. The country has to deal with the problem of declining soil fertility and the constraints associated with climate variability through investment in appropriate irrigation, and soil conservation, and rehabilitation programmes.

The close link between development and environment is well recognised but less manifested in development policies. The Ethiopian government's 2010–2015 Five Year Growth and Transformation Plan promises to build 'a "green" and climate change economy.' But the plan provides far less comprehensive action plans to rehabilitate the country's degraded environment other than proposing a limited expansion of forest cover and parks over the plan period. The Plan primarily focuses on building infrastructure, education, and export-led economy that is aimed at producing a middle class, and environmental consideration does not figure significantly in the planned development activities. Much of the Ethiopian landscape has been significantly altered and reshaped by centuries of socio-economic activities. Currently, nearly three-quarters of the rural population are living on degraded land.[147] Given the heavy ecological toll that the country has suffered over a long period of its history, environmental concerns should be at the forefront of the government's development policy-making.

Notes

1 James Newman. 1995. *The Peopling of Africa*, New Haven: Yale University Press, p. 90.
2 Karl Butzer. 1981. Rise and fall of Axum, Ethiopia: a geo-archaeological interpretation. *American Antiquity* 46, 3 (July): 471–95; p. 472.
3 Andrew Goudie. 2000. *The Human Impact on the Natural Environment*, Cambridge, MA: Cambridge University Press, p. 160.
4 For a detailed discussion of soil-forming factors and soil characteristics, composition and profile, see Oliver S. Owen and Daniel D. Chiras. 1990. *Natural Resource Conservation: An Ecological Approach*, New York: McMillan Publishing Co., pp. 59–77; S. Ellis and A. Miller. 1995. *Soils and Environment*, New York: Routledge, pp. 96–110.
5 Duane L. Winegardner. 2008. The fundamental concept of soil. In: *Environment: An Interdisciplinary Anthology*, Glenn Adelson, James Engell, Brent Ranalli, and K. P. Van Anglen (eds.), New Haven: Yale University Press, pp. 415–8; p. 416.
6 Berhanu Debele. 1989. The role of land use planning in Ethiopia's national food strategy, Towards a Food and Nutrition Strategy in Ethiopia. *The Proceedings of the National Workshop on Food Strategies for Ethiopia*, Haromaya University, 8–12 December, 1986, pp. 180–205.
7 Ethiopian Mapping Authority. 1988. *National Atlas of Ethiopia*, Addis Ababa: Ethiopian Mapping Authority, p. 8.
8 Mesfin Abebe. 1998. *Nature and Management of Ethiopian Soils*, Haramaya: Haramaya University Press, p. 196.

9 Ibid., pp. 197–8; Tilahun Amede, Takele Belachew and Endrias Geta. 2001. *Reversing the Degradation of Arable Land in the Ethiopian Highlands*. Managing Africa's Soils No. 23, IIED, p. 4; Paulos Dubale. 2001. Soil and water resources and degradation factors affecting productivity in Ethiopian highland agro-ecosystems, *Northeast African Studies* 8, 1: 27–52.
10 Ethiopian Mapping Authority, p. 8.
11 Mesfin Abebe, p. 196.
12 Samuel C. Jutzi. 1989. The Ethiopian vertisols: a vast natural resource, but considerably underutilized. In: *First Natural Resources Conservation Conference, Natural Resources Degradation: A Challenge to Ethiopia*, February 1–8, 1989, Addis Ababa, pp. 41–45; p. 41.
13 Desta Beyene. 1988. Soil fertility research on some Ethiopian vertisols. In: *Proceedings of a Conference on Management of Vertisols in Sub-Saharan Africa held at ILCA*, Addis Ababa, 31 August–4 September 1987, pp. 223–31; Jutzi, p. 42.
14 Asnakew Wolde-Ab. 1987. Physical properties of Ethiopian vertisols. In: *Proceedings of a Conference on Management of Verisols in Sub-Saharan Africa held at ILCA*, Addis Ababa, 31 August–4 Septemeber 1987, pp. 111–23; Mesfin Abebe, p. 161.
15 Mesfin Abebe, p. 161.
16 Ibid., pp. 161–2.
17 Ibid., p. 87–9; Paulos Dubale, pp. 29–35.
18 Mesfin Abebe. p. 87–9.
19 Ibid., p. 174.
20 Ibid., p. 177.
21 Ibid., pp. 177–8.
22 ISRIC, International Soil Reference and Information Centre, The Netherlands. Accessed online at www.isric.org/ISRIC/webdocs/docs/major_soils_of_the_world/annexes/diag_hor.pdf.
23 Ethiopian Mapping Authority, p. 7.
24 Ibid., p. 8.
25 Ibid., pp. 7–8; www.fao.org/ag/agl/agll/wrb/wrbmaps/htm/acrisols.htm.
26 Ethiopian Mapping Authority, pp. 7–8.
27 Ayenew Haileselassie. 2004. Ethiopia's struggle over land reform, *World Press Review*, 51, 4. Available online at www.worldpress.org/Africa/1839.cfm (accessed 17 August 2011).
28 Melaku Worede, Tesfaye Tesemma, and Regassa Feyissa. 1999. Chapter 6. Keeping diversity alive: an Ethiopian perspective. In: *Genes in the Field: On-Farm Conservation of Crop Diversity*, Stephen B. Bush (ed.), London: Lewis Publishers, pp. 1–6; p. 1.
29 Ibid., pp. 1–2.
30 Ibid., p. 2.
31 United Nations Development Program. 2013. *Human Development Report* 2013, New York: UNDP, Table 13, p. 192.
32 Ibid., p. 4; Plant Genetic Resources Center/Ethiopia (PGRC). 1986. *Ten Years of Collection, Conservation and Utilization*, 1976–1986, Addis Ababa: PGRC, p. 12.
33 Woody Biomass Inventory and Strategic Planning Project (WBISPP). 1995. *Annex 3: Socio-Cultural and Economic Aspects of Crop, Livestock and Tree Production*, WBISPP, November 30, p. 23.
34 Samuel Gebreselassie. 2006. *Land, Land Policy and Smallholder Agriculture in Ethiopia: Options and Scenarios*. Paper prepared for the Future Agricultures Consortium meeting at the Institute of Development Studies, 20–22 March 2006, Addis Ababa: Institute of Development Research, p. 13.
35 CSA. 2009. *Report on Land Utilization*, Volume IV, Statistical Bulletin No. 446, Addis Ababa: CSA, Table 1, p. 13.
36 Mengistu Geza. 1999. Harnessing techniques and work performance of draught horses in Ethiopia. In: *Meeting the Challenges of Animal Traction*. A resource book of the

Animal Traction Network for Eastern and Southern Africa (ATNESA), P. Starkey and P. Kaumbutho (eds.), Harare, Zimbabwe: Intermediate Technology Publications, pp. 144–7; p. 144. The number of draught oxen per smallholder is only 1.02: CSA. 2003. *Ethiopian Agricultural Sample Enumeration: Statistical Report on Farm Management Practices, Livestock and Farm Implements*, Addis Ababa: CSA. Cited in Samuel Gebreselassie. 2006. *Land, Land Policy and Smallholder Agriculture in Ethiopia: Options and Scenarios*. Paper prepared for the *Future Agricultures Consortium meeting at the Institute of Development Studies*, 20–22 March, Addis Ababa: Institute of Development Research, p. 13.

37 Mengistu Geza, Ibid.; WBISPP, p. 23.

38 Michael Mortimore. 1991. *A Review of Mixed Farming System in the Semi-Arid Zone of Sub-Saharan Africa*, Addis Ababa: International Livestock Center of Africa, p. 8.

39 Ibid.

40 WBISPP, pp. 82–3.

41 Ibid., pp. 90–1.

42 National Research Council. 2009. *Emerging Technologies to Benefit Farmers in Sub-Saharan Africa and South Asia*, Washington, DC: The National Academies Press, p. 39.

43 Demel Teketay. 1999. *Deforestation, Wood Famine, and Environmental Degradation in Ethiopia's Highland Ecosystems: Urgent Need for Action*, Forest Stewardship Council (FSC Africa), Kumasi, Ghana, pp. 61–9; pp. 72–5.

44 Ibid., p. 91.

45 Ibid., p. 92.

46 David Bourn. 2002. *Farming in Tsetse Controlled Areas of Eastern Africa Ethiopia National Component: Farming Systems and Natural Resource Management*. Short Term Technical Assistance Consultancy Report, Project 7 ACP ET086, Addis Ababa, Ministry of Agriculture, p. 7.

47 WBISPP, p. 83.

48 CSA. 2010. *Ethiopia: Statistical Abstract 2009*, Addis Ababa: CSA.

49 Ministry of Water Resources (MoWR) and National Meteorological Services Agency (NMSA). 2001. *Initial National Communication of Ethiopia to the United Nations Framework Convention on Climate Change (UNFCCC)*, Addis Ababa, Ethiopia, p. 32.

50 Bourn, pp. 16–7.

51 Ministry of Agriculture. 2014. *ZemenawiGibrinatchen*, vol. 3, No. 1, Yekatit 2006 E.C., p. 7.

52 Ministry of Water Resources. 2006. *Five Year Irrigation Development Program: 2005/2006–2009/2010*. Addis Ababa, Ethiopia: MoWR.

53 S. B. Awulachew, M. Loulseged, and A. D. Yilma (eds.). 2007. Impact of irrigation on poverty and environment in Ethiopia. In: *Proceedings of the Symposium and Exhibition*, Ghion Hotel, Addis Ababa, Ethiopia. November 27–29.

54 McKee, p. 65.

55 Ibid.

56 S. B. Awulachew. 2010. *Irrigation Potential in Ethiopia: Constraints and Opportunities for Enhancing the System*. Cited in Seleshi Bekele Awulachew and Mekonnen Ayana. 2010. Performance of irrigation: an assessment at different scales in Ethiopia, *Experimental Agriculture*, 47, 1: 57–69, p. 62. Available online at http://journals.cambridge.org/action/displayFulltext?type=1&fid=7967444&jid=EAG&volumeId=47&issueId=S1&aid=7967442 (accessed 30 March 2011).

57 Fitsum Hagos, G. Makombe, R. E. Namara, and Seleshi Bekele Awulachew. 2009. *Importance of Irrigated Agriculture to the Ethiopian Economy: Capturing the Direct Net Benefits of Irrigation*. Colombo, Sri Lanka: International Water Management Institute, IWMI Research Report 128, p. 7; The World Bank. 2006. *Managing Water Resources to Maximize Sustainable Growth*, Ethiopia, Report 36000-ET, World Bank: Washington, DC, pp. 21–2.

58 Ministry of Water Resources. 2010. *Irrigation and Drainage Projects in Ethiopia*, Addis Ababa: Ministry of Water Resources. Available online at www.mowr.gov.et/index. php?pagenum=4.2 (accessed 6 October 2010).

59 Amare Haileslassie, Fitsum Hagos, Everisto Mapedza, Claudia Sadoff, Seleshi Bekele Awulachew, Solomon Gebreselassie, and Don Peden. 2008. *Institutional Settings and Livelihood Strategies in the Blue Nile Basin: Implications for Upstream/ Downstream Linkages*, Working Paper 132, International Livestock Research Institute (ILRI) and International Water Management Institute, p. 22. Available online at www. indiaenvironmentportal.org.in/files/WOR132.pdf (accessed 5 October 2010).

60 Michael Kugelman and Susan L. Levenstein. 2013. *The Global Farms Race: Land Grabs, Agricultural Investment, and the Scramble for Food Security*, London: Island Press, p. 71.

61 ILO, p. 3.

62 The World Bank. 2007. *Rising Global Interest in Farmland: Can It Yield Sustainable and Equitable Benefits?* Washington, DC: The World Bank, p. 44. Some 350,000 hectares of land has been leased to foreign investors since 2008 in southwest Ethiopia alone. Jenny Vaughan, Ethiopia land lease risks displacement, *AFP*, July 29, 2011. Available online at www.oaklandinstitute.org/ethiopia-land-lease-risks-displacement-report (accessed 29 June 2016).

63 Atakilte Beyene. 2011. Smallholder-led transformation towards biofuel production in Ethiopia. In: *Biofuels, Land Grabbing and Food Security in Africa*, Prosper B. Matondi, Kjell Havnevik and Atakilte Beyene (eds.), London: Zed Books, pp. 90–105; p. 104.

64 Fred Pearce. 2012. *The Land Grabbers: The New Fight Over Who Owns the Earth*, Boston: Beacon Press, pp. 5–6.

65 Ibid., p. 13.

66 The Economist. 2009. *Outsourcing's Third Wave*, 21 May. Available online at www. economist.com/node/13692889 (accessed 21 December 2012). Cited in Michael Kugelman and Susan L. Levenstein. 2013. *The Global Farms Race: Land Grabs, Agricultural Investment, and the Scramble for Food Security*, London: Island Press, p. 14.

67 David Hallam. 2013. Overview. In: *The Global Farms Race: Land Grabs, Agricultural Investment, and the Scramble for Food Security*, Michael Kugelman and Susan L. Levenstein (eds.), London: Island Press, p. 50.

68 Paul Richardson. 2012. Ethiopia to accelerate land commercialization amid opposition. Available online at www.bloomberg.com/news/articles/2012-03-23/ethiopia-to-accelerate-land-commercialization-amid-opposition (accessed 29 June 2016).

69 Ministry of Agriculture. 2014. *ZemenawiGibrinatchen*, Vol. 3, No. 1, Yekatit 2006 E.C., p. 15.

70 The Oakland Institute. 2011. *Understanding Land Investment Deals in Africa, Country Report: Ethiopia*, Oakland: The Oakland Institute.

71 Pearce, p. 12.

72 Michael Taylor. 2010. Quoted in John Vidal. 2010. Billionaires and Mega-Corporations Behind Immense Land Grab in Africa. *Mail and Guardian*, 10 March, p. 5.

73 Alexandra Spieldoch and Sophia Murphy. 2013. Social and economic implications. In: *The Global Farms Race: Land Grabs, Agricultural Investment, and the Scramble for Food Security*, Michael Kugelman and Susan L. Levenstein (eds.), London: Island Press, pp. 66–7. Other international organisations, such the Food and Agriculture Organization (FAO), the International Fund for Agricultural Development (IFAD), the United Nations Conference on Trade and Development (UNCTAD), and the World Bank, have also initiated similar guiding principles for socially and environmentally responsible agricultural investments in developing countries. These principles include: '1) Respect for land and resources rights. Existing rights to land and natural resources are recognized and respected. 2) Food security and

rural development. Investments do not jeopardize food security and rural development, but rather strengthen it. 3) Transparency, good governance, and enabling environment. Processes relating to investment in agriculture are transparent and carefully monitored, ensuring accountability by all stakeholders. 4) Consultation and participation. All those materially affected are consulted, and agreements from consultation are recorded and enforced. 5) Economic viability and responsible agro-enterprise investing. Projects are viable economically, respect the rule of law, reflect industry best practices, and result in durable shared value. 6) Social sustainability. Investments generate desirable social and distributional impacts, and do not increase vulnerability. 7) Environmental sustainability. Environmental impacts are quantified, and measures are taken to encourage sustainable resource use while minimizing and mitigating negative impacts.' See David Hallam. 2013. Overview. In: *The Global Farms Race: Land Grabs, Agricultural Investment, and the Scramble for Food Security*, Michael Kugelman and Susan L. Levenstein (eds.), London: Island Press, p. 55–6).

74 World Bank. 2012. Inclusive green growth, *World Bank Research Digest*, 6, 4 (Summer), pp. 4–5.
75 Vidal, p. 5.
76 Booming floriculture industry fuels Ethiopian economic growth. *The Guardian*, 14 February 2008. Available at http://www.africanagriculture.co.zw/2008/02/booming-floriculture-industry-fuels.html (accessed 6 July 2016).
77 Ibid.
78 United Nations. 2012. COMTRADE.
79 Zelalem T. Chala. 2010. *Economic Significance of Selective Export Promotion on Poverty Reduction and Inter-Industry Growth of Ethiopia*. PhD dissertation, Virginia Polytechnic Institute and State University, p. 13.
80 UN, COMTRADE.
81 *Jimma Times*. 19 February 2008. Environmental, labour concerns grow over Ethiopian floriculture industry. Available online at http://africanagriculture.co.zw /2008/02/environmental-labor-concerns-grow-over.html (accessed 9 July 2016).
82 Abreham Berta and Belay Zerga. 2015. Biofuel energy for mitigation of climate change in Ethiopia. *Journal of Energy and Natural Resources* 4, 6 (December), pp. 62–72. Available at www.articlepublishinggroup.com/html/10/11648.j.jenr.20150406.11.html (accessed 9 July 2016).
83 Ministry of Mines and Energy of Ethiopia (MoME). 2007. *The Biofuel Development and Utilization Strategy of Ethiopia*, Addis Ababa, Ethiopia.
84 Atakilte Beyene. 2011. Smallholder-led transformation towards biofuel production in Ethiopia. In: *Biofuels, Land Grabbing and Food Security in Africa*, Prosper B. Matondi, Kjell Havnevik and Atakilte Beyene (eds.), London: Zed Books, pp. 90–105, p. 94.
85 Wudineh Zenebe. 2007. Ethiopia's New Energy Strategy Opens Way for Biofuel. *Addis Fortune*, 4 November.
86 Kassaye Tekola Moges. 2010. *Smallholder Farmers and Biofuel Farmers' Perspectives in Growing Castor Beans in Ethiopia*. Master's thesis, Uppsala, Swedish University of Agricultural Sciences, p. 8. Vermeulen Dufey. 2007. *Biofuels Strategic Choices for Commodity Dependent Developing Countries*. Common Fund for Commodities (CFC), Amsterdam, The Netherlands.
87 Kassaye, pp. 7–8.
88 N. H. Ravindranath, Ritumbara Manuvie, and C. Sita Lakshmi. 2010. Biofuels and climate change. In: *Food versus Fuel: An Informed Introduction to Biofuels*, Frank Rosillo-Calle and Francis X. Johnson (eds.), London: Zed Press, pp. 138–63; pp. 153–4.
89 A. Eide. 2008. *The Right to Food and the Impact of Liquid Biofuels (Agrofuels)*. Right to Food Studies. Rome: Food and Agriculture Organization.

90 Ibid.
91 L. Engstrom. 2009. *Liquid Biofuels – Opportunities and Challenges in Developing Countries*: A Summary Report from SIDA's Helpdesk for Environmental Assessment, May.
92 Ibid.
93 Kassaye, p. 10.
94 CSA. 2010. *Statistical Abstract 2009*, Addis Ababa: CSA, p. 44.
95 Ministry of Water Resources (MoWR) and National Meteorological Services Agency (NMSA). 2004. *Annual Climate Bulletin Year 2004*, Addis Ababa: NMSA.
96 Environmental Protection Authority. 1998. *National Action Program to Combat Desertification*, Addis Ababa: EPA, p. 14.
97 Girma Kebbede. 1992. *The State and Development in Ethiopia*, Atlantic Highlands, NJ: Humanities Press, pp. 66–7.
98 Jonathan McKee. 2007. *Ethiopia: Country Environmental Profile, 2007*, pp. 25–6.
99 C. McSweeney, M. New, and G. Lizcano. 2008. *UNDP Climate Change Country Profiles – Ethiopia*; Cited in *Climate-Related Vulnerability and Adaptive-Capacity in Ethiopia's Borana and Somali Communities*, Béatrice Riché, Excellent Hachileka, Cynthia B. Awuor, and Anne Hammill. Final assessment report. Save the Children (UK). 2009. Available online at www.iisd.org/pdf/2010/climate_ethiopia_communities.pdf, 'p. 22 (accessed 4 November 2010).
100 Ibid.
101 Ibid.
102 The wet years included 1958, 1961, 1964, 1967, 1968, 1977, 1993, 1996, 1998 and 2006. The dry years were 1952, 1959, 1965, 1972, 1973, 1978, 1984, 1991, 1994, 1999, and 2000.
103 Ethiopian NAPA. 2007. *Climate Change National Adaptation Program of Action*, Addis Ababa, Ethiopia; McSweeney and Lizcano, p. 22.
104 J. C. Nkomo, A. O. Nyong, and K. Kulindwa. 2006. *The Impacts of Climate Change in Africa*. The Stern Review on the Economics of Climate Change, p. 26.
105 UN Office for the Coordination of Humanitarian Affairs. 2006. *Ethiopia: Dire Dawa Floods*, Ref. OCHA/GVA-2006/0143.
106 National Metreological Services Agency (NMSA). 2006. *Agro-meteorology Bulletin*, Addis Ababa: NMSA.
107 Aklilu Amsalu and Alebachew Adem. 2009. *Assessment of Climate Change-induced Hazards, Impacts and Responses in Southern Lowlands of Ethiopia, Forum for Social Studies Research*. Report No. 4, Addis Ababa: Forum for Social Studies, p. 23.
108 The International Research Institute for Climate and Society, IRI Scientist Wins NSF CAREER Award, October 15, 2010. Unpublished Document.
109 United Nations Educational, Scientific, and Cultural Organization World Water Assessment Program. 2004. *National Water Development Report for Ethiopia*, Final Report, Addis Ababa, Ministry of Water Resources, December, UN-WATER/WWAP/2006/7, p. 71. Available online at http://unesdoc.unesco.org/images/0014/001459/145926e.pdf (accessed 26 November 2010).
110 F. J. Meza, J. W. Hansen, and D. Osgood. 2008. Economic value of seasonal climate forecasts for agriculture: review of ex-ante assessments and recommendations for future research. *Journal of Applied Meteorology and Climatollgy*, 47, 1269–1286; Y. Abawi, P. Llanso, M. Harrison, and S. J. Mason, 2008. Water, health and early warnings. In: *Seasonal Climate: Forecasting and Managing Risk*. A.Troccoli, M. Harrison, D. L. T. Anderson, S. J. Mason (eds). Dordrecht: Springer, pp. 351–95. DOI: 10.1007/978-1-4020-6992-5-13; S. J. Connor, M. C. Thomson, and B. Menne. 2008. A multimodel framework in support of malaria surveillance and control. In: *Seasonal Forecasts, Climatic Change and Human Health. Health and Climate*

Series: Advances in Global Change Research, M. C. Thomson, R. Garcia-Herrera, and M. Beniston (eds.), 30, DOI: 10.1007/978-1-4020-6877-5. Cited in Carlos A. Nobre. 2009. How can current and future early warning systems be used to enhance adaptive capacity to climate change? *Development in a Changing World*, Blogs.WorldBank. org, October 5. Available online at https://blogs.worldbank.org/climatechange/how-can-current-and-future-early-warning-systems-be-used-enhance-adaptive-capacity-climate-change (accessed 27 October 2010).

111 Ibid., p. 133.

112 Celia A. Harvey and David Pimentel. 1996. Effects of soil and wood depletion on biodiversity. *Biodiversity and Conservation* 5: 1121–30; p. 1121.

113 Demel Teketay, Masresha Fetene, and Asferachew Abate. 2003. The state of the environment in Ethiopia: past, present, and future prospects. In: *Environment and Environmental Change in Ethiopia*, Gedion Asfaw (ed.), Addis Ababa: Forum for Social Studies.

114 John Campbell. 1991. Land or peasants? The dilemma confronting Ethiopian resource conservation. *African Affairs* 90: 5–21; p. 7.

115 Lulseged Tamene and Paul L. G. Vlek. 2007. Soil erosion studies in northern Ethiopia. In: *Land Use and Soil Resources*, Ademola K. Braimola and Paul L. G. Vlek (eds.), Stockholm: Swedish Academy of Sciences, p. 73; pp. 73–100.

116 Abdalla A. Ahmed and Hamid A. E. Ismael. 2008. Sediment in the Nile River system, Khartoum: UNESCO International Sediment Initiative. Available online at www.irt-ces.org/isi/isi_document/Sediment%20in%20the%20Nile%20River%20System.pdf (accessed 17 January 2012).

117 WBISPP, p. 126.

118 Z. G. Bai, D. L. Dent, L. Olsson, and M. E. Schaepman. 2008. *Global Assessment of Land Degradation and Improvement: 1. Identification by Remote Sensing*, Wageningen, The Netherlands, World Soil Information, Report 2008/01ISRIC, p. 25.

119 Paulos Dubale. 2001. Soil and water resources and degradation factors affecting productivity in Ethiopian highland agro-ecosystems. *Northeast African Studies* 8, 1: 27–52.

120 W. Gete. No date. *Natural Resources, Poverty, and Conflict in Sub-Saharan Africa: Evidence from Ethiopia*, pp. 18–25.

121 Markos Ezra. 1997. *Demographic Responses to Ecological Degradation and Food Insecurity: Drought Prone Areas in Northern Ethiopia*. PhD dissertation, The Netherlands Graduate School of Research and Demography, p. 63; WBISPP, p. 53.

122 R. Barber. 1984. *An Assessment of the Dominant Soil Degradation Process in the Ethiopian Highlands: Their Impacts and Hazards*, Addis Ababa: Ministry of Agriculture, Land Use Planning and Regulatory Department.

123 Girma Kebbede, pp. 58–60.

124 Demel Teketay. 1999. *Deforestation, Wood Famine, and Environmental Degradation in Ethiopia's Highland Ecosystems: Urgent Need for Action*, Forest Stewardship Council (FSC Africa), Kumasi, Ghana, pp. 61–9; pp. 72–5.

125 Million Alemayehu. 2000. Indigenous conservation practices in North Shewa Administrative Zone, Amhara National Regional State, environment and development in Ethiopia. In: *Proceedings of the Symposium of the Forum for Social Studies*, Zenbework Tadesse (ed.), Addis Ababa, 15–16 September, pp. 92–8.

126 Kebede Tato. 1990. Ethiopian soil conservation program and its future trends. In: *National Conservation Strategy Conference Document*, Vol. 2, Introduction and Africa's Experience of National Conservation Strategies, Addis Ababa: ONCCP; WBISPP, pp. 125–65; Tilahun Amede, Takele Belachew and Endrias Geta. 2001. *Reversing the degradation of arable land in the Ethiopian Highlands*. Managing Africa's Soils No. 23, International Center for Research in Agroforestry, May.

127 G. Tyler Miller and Scott E. Spoolman. 2002. *Living in the Environment*, Belmont, CA: Wadsworth Group, G3.

128 WBISPP, pp. 16–78.

129 Tilahun Amede et al., p. 12.

130 Belay Tegene. 1998. Indigenous soil knowledge and fertility management practices of the south Wello Highlands. *Journal of Ethiopian Studies* 32: 123–58.

131 Adam Szirmai. 2005. *The Dynamics of Socio-Economic Development*, Cambridge: Cambridge University Press, p. 368.

132 WBISPP, p. 130.

133 Ellis and Mellor, pp. 229–30.

134 Paul Harrison. 1987. *The Greening of Africa: Breaking through in the Battle for Land and Food*, London: Penguin, p. 75.

135 WBISPP, p. 130.

136 Carlos Oya. 2011. Agro-pessimism, capitalism and agrarian change: trajectories and contradictions in Sub-Saharan Africa. In: *The Political Economy of Africa*, Vishu Padayachee (ed.), pp. 85–109; See graph, p. 96.

137 H. W. O. Okoth-Ogendo. 1993. Agrarian reform in Sub-Saharan Africa: an assessment of state responses to the african agrarian crisis and their implications for agricultural development. In: *Land in African Agrarian Systems*, Thomas J. Bassett and Donald E. Crummey (eds.), Madison: The University of Wisconsin Press, p. 256; pp. 247–73.

138 Gebru Mersha and Mwangi wa Gĩthĩnji. 2005. *Untying the Gordian Knot: The Question of Land Reform in Ethiopia*. Land, Poverty and Public Action Policy Paper No. 9, Institute of Social Studies/United Nations Development Program, New York, pp. 21–6; p. 23.

139 Teferi Abate. 1995. Land redistribution and inter-household relations: the case of two communities in northern Ethiopia. *Ethiopian Journal of Development Research* 17: 23–40; J. P. Sutcliffe. 1995. Soil conservation and land tenure in highland Ethiopia, Special Issue on Land rights and Access to Land in Post-Derg Ethiopia. *Ethiopian Journal of Development Research* 17: 63–80; Dessalegn Rahmato. 1999. Revisiting the land issue: options for change, *Economic Focus*, 2, 4 (August).

140 Dessalegn Rahmato. 1999. Revisiting the land issue: options for change, *Economic Focus*, 2, 4 (August); Dessalegn Rahmato. 2009. *The Peasant and the State: Studies in Agrarian Change in Ethiopia 1950s–2000s*, Addis Ababa: Addis Ababa University Press, pp. 185–9.

141 Federal Democratic Republic of Ethiopia. 2010. *Growth and Transformation Plan, 2010/11–2014/15*, Addis Ababa, p. 9.

142 Ministry of Agriculture. 2014. *ZemenawiGibrinatchen*, vol. 3, No. 1, Yekatit 2006 E.C., p. 5.

143 Gebru Mersha and Mwangi wa Gĩthĩnji, p. 25.

144 E. C. Oerke, H. W. Dehne, F. Schonbeck, and A. Weber. 1994. *Crop Production and Prop Protections. Estimated Losses in Major Food and Cash Crops*, Amsterdam: Elsevier. Quoted in National Research Council. 2009. *Emerging Technologies to Benefit Farmers in Sub-Saharan and South Asia*, Washington, DC: The National Academies Press, p. 44.

145 Zerihun Gudeta. 2009. How successful the Agricultural Development-Led Industrialization Strategy (ADLI) will be leaving the existing landholding system intact? A major constraint for the realization of ADLI's target. *Ethiopian e-Journal for Research and Innovation Foresight* 1, 1 (December), p. 3; Mulat Demeke, Ali Said, and T. S. Jayne. 1997. Promoting fertilizer use in ethiopia: the implication of improving grain market performance, input market efficiency, and farm management,

Working Paper 5, Grain Market Research Project, Ministry of Economic Development and Cooperation, Addis Ababa.

146 Yeraswork Yilma. 1997. Integrated Pest Management (IPM) Experience in the Amhara Region. In: *Proceedings of Integrated Pest Management Workshop in the Amhara Region*, 24–26 February, Dessie, 28–33.

147 United Nations Development Program. 2013. *Human Development Report 2013*, New York: UNDP, Table 13, p. 192.

3 Pastoral resources

Pastoralism – a way of life based on keeping herds of grazing animals – is an extensive form of land use in Ethiopia. Ethiopian lowland areas, which typically have elevations of 1,500 metres (m) or less, constitute rangelands and are found in the northeastern, eastern, southern, and southwestern parts of the country. Nine of ten people living in these areas pursue pastoral and agropastoral activities as primary modes of livelihood. Many of these areas lie in the semi-arid and arid zones of the country. Precipitation is low and unpredictable and, compared to the highlands, the lowlands are at an increased risk of drought. Rangelands occupy more than one-half of the country's total land area and provide a livelihood for about 10 per cent of the country's population and home for 40 per cent of the estimated cattle population, 75 per cent of the goats, 25 per cent of the sheep, 20 per cent of the equines, and 100 per cent of the camels.[1] This chapter explores Ethiopia's pastoral resources – their ecology, distribution, and condition.

Pastoralism has been practised extensively for centuries throughout large portions of Ethiopia, especially in the Borana Zone of the Oromiya regional state, in the Somali and Afar regional states, and in parts of the Gambela, SNNP, and Benshangul-Gumuz regional states. Major Ethiopian pastoralist groups include the Borana of the south, the Afar of the northeast, and Somali of the southeast. There are also smaller agropastoral populations that live in the southwest. The types of animals each group raises vary, depending on the environment. For Borana pastoralists, it is primarily cattle. Among the Afar, it is a mixture of cattle, goats, camels, and sheep. The Somali prefer camels, although ruminants are increasingly more common in recent years. Ethiopian pastoralists share common livelihood characteristics despite differences in environment and culture. Because the environments in which they live are highly variable and unreliable, they all have developed adaptive mechanisms to withstand environmental hardships. These adaptive mechanisms include: mobility, breeding livestock resistant to drought and disease, diversifying herd stock, augmenting herd size during good years, and ensuring communal resources benefit all without damaging them, if possible. Ethiopian pastoralists live in the lowlands on the periphery of the central, more densely populated highlands and are among the most marginalised communities. They access fewer infrastructure, methods of communication, basic social services such as clean water, clinics, schools, security, and administrative centres, and are often somewhat removed from markets and urban centres.[2]

Pastoral livelihoods in Afar regional state

The Afar regional state is home to over one-quarter of the Ethiopian pastoralists.[3] The region is located in northeastern Ethiopia and covers a little more than 72,000 square kilometres. This triangular region of lowland plains lies within the Ethiopian Rift Valley, bounded in the west by the eastern escarpment of the northern highlands and on the south by the northern escarpment of the Hararghe highlands. Lava and sand plains, which are at times below sea level, characterise much of the landscape. Elevations range from 116 m below sea level to 2,063 m at Mount Mussa-Alle. The region is dominated by arid conditions. This aridity is a major limiting factor for resource use. Rainfall in the region occurs in two separate seasons. The season of small rains lasts through March and April and the season of more substantial rains lasts through July, August, and September. Mean annual rainfall ranges from 400 millimetres (mm) around the Gewane to 200–300 mm at Mille and 200 mm in Assaita. These figures can be misleading because spatial and temporal irregularities characterise rainfall in this part of the country. Temperatures are very high in the region. The highest temperatures are experienced during the months of June, July, and August. During this period, the minimum monthly temperatures stay around 25 degrees centigrade at higher altitudes and 35 degrees centigrade or more at lower altitudes. Much of the state is highly inhospitable due to its aridity. The northern half of the state has scarce vegetation. The Awash River, which starts in the central highlands, southwest of Addis Ababa, is the only permanent source of water other than wells. In the semi-arid areas, vegetation varies from a dense thicket to open shrubland to grassland with no woody species. Along the middle valley, riparian forests cover the river-banks downstream. The dominant species there is acacia *nilotica*, which forms a closed canopy. The understory tends to be open, allowing for grazing by live-stock.[4] As one moves away from the Awash River, vegetation fades into woodland and savanna. Further away to the north and east, vegetation dissipates into a mere sub-desert steppe, and then finally into the only scant shrubbery.[5]

The vast majority of the state's population of 1.5 million is ethnic Afar. The Afar are primarily pastoralists. Their economy has adapted to the unusually harsh environment. Like many pastoralists, they keep herds composed of multiple species (cattle, camel, goats, and sheep) that fulfil different roles, such as providing milk and meat for consumption and trade. The objective of this system is to maintain stock levels while providing goods for domestic use, social exchange, and sale on the market. The subsistence base depends on mobility, which allows herders to use natural resources that vary seasonally and geographically. The greatest challenge in a pastoralist economy is managing the livestock population and preventing losses caused by factors such as drought or disease. Institutions that regulate the exchange of livestock and labour meet this challenge.[6]

While the Afar may practise other economic activities, such as trade and small-scale cultivation, subsistence pastoralism is the main mode of survival. Pastoral values predominate in social and cultural life. Of the species kept, none is valued more than a camel. The Afar herders practise transhumant pastoralism, which involves both short- and long-range mobility. Mobility is a key strategy in preventing livestock losses, for it allows pastoralists to take advantage of grazing and water resources in different areas. The Afar herders typically travel between the dry season grazing grounds along the Awash River and wet-season grazing grounds such as the Alledeghi Plains. Short-range mobility involves dividing the herd into separate units. Livestock is often separated into lactating or dry herds. These herds are further divided into units of camels, cattle, goats, and sheep. The lactating herd is kept near the main settlement, while the dry herd is sent out into satellite encampments.[7]

The Afar are not the sole inhabitants of the Awash River Valley. They have several other pastoralists, agropastoralists, and agriculturalist groups as neighbours. These include: the Argoba and Tigray cultivators to the north, the Issa-Somali pastoralists to the east, and the Karrayu, Jille, Ittu, and Harsu Oromo to the southwest. Inter-tribal conflicts are commonplace between the Afar and these neighbouring communities. The tension between the Issa-Somali – the most numerous of the Afar's neighbours – and the Afar are age-old and are further exacerbated by the escalating water and pasture shortages. The Issa-Somali occupy the arid parts of northeastern Ethiopia, especially the Alledeghi Plain. Like the Afar, they are primarily pastoralists inhabiting the same general geographic area. The two groups frequently conduct raids against one another. Competition over scarce resources, especially over access to the wet-season pastures of the Alledeghi Plain and the dry-season pastures along the Awash valley, aggravates the conflict. The Issa Somali have been gradually expanding from their homeland in the east northward and westward into Afar land, pushing the Afar into the Awash flood plain, which has been the land of the Karrayu, Ittu, Jille Oromo, and the Argoba.[8] The Karrayu are also traditional enemies of the Afar and conflicts over water and land are not uncommon. The inter-tribal conflicts are closely connected with the rains. When the rains are good, grass is abundant, and herders do not need to travel far from their home pastures to forage. During times of drought, however, herders are often forced to go further afield in search of grazing, which brings them into conflict with other pastoralists pursuing the same goal.[9]

The Awash River Basin covers an area of about 123,400 square kilometres. The basin is divided into the upper valley, middle valley, and lower valley or Plains. The Afar inhabit most of the middle valley and almost the entire lower valley. The flat plains of the Awash valley contain the most fertile soils in Ethiopia. The plains are nourished annually by extensive flooding that heavy rains in the highlands generate during the wet season. During the rainy season, the Awash alluvial plains are completely inundated and, therefore, unsuitable for grazing or farming. The flood plains gradually dry out during the dry season, providing enough grazing for the entire eight months of dryness.[10]

The Afar nomadic pastoral system was once largely in balance with available resources. The Afar maintained a sustainable economy through multi-species livestock keeping, relying on dairy cows, goats, sheep, and camels as food sources. However, over the last half century, much of this resource base has been lost to other uses and, as a result, the Afar have come under great pressure, and their way of life has been threatened. The land tenure system in the Afar region was communal, and access to community-owned land was a birthright, with land being passed down from father to son. Pastureland was apportioned among clans and sub-clans based on customary rules. Access to and use of other clan's grazing pastures required permission. They shared clan resources during times of hardship. However, in 1955, the revised constitution of Haile Selassie's government dispossessed all pastoralists of their land and declared it state property.[11] With the construction of the Koka Dam in the 1950s, the area took a new economic significance to the central government. This dam generated power, provided sufficient irrigation water year-round, almost eliminated flooding in the upper valley, and moderated seasonal floods in the middle and lower plains. In 1961, the first large irrigation scheme was set up at Tendaho in the lower Awash plains. In 1962, the Awash Valley Authority (AVA) was established to administer and legally superintend all development projects in the area. Most development projects took the form of large, mechanised commercial enterprises under the management of the AVA. Foreign agro-industrialists in joint ventures with the state ran many of the farm projects. By the mid-1990s, nearly 70,000 hectares of prime pasture were lost to various large- and medium-scale irrigation schemes, including Metehara, Abadir, and Amibara-Malka-Saddi.[12]

The clearing of huge areas of riverine vegetation to make room for irrigation farms resulted in the decrease of sizeable amounts of dry-season fodder reserves. As a result, the Afar were forced to concentrate their animals in the limited wet-season areas, leading to localised overgrazing, destruction of the natural vegetation, and increased soil erosion.[13] This, in turn, has destroyed the grazing lands and has led to a decline in livestock production. The irrigation projects have also blocked the migration routes of the Afar, hence denying them access to the traditional livestock watering holes. Furthermore, the introduction of large irrigation systems has affected the quality of the soils in the Afar region. Soil salinity problems have been identified in all the irrigation schemes in the Awash valley, forcing the abandonment of previously irrigated lands and expropriation of more pastureland.[14] Although drainage projects are in place to leach out the excess salts, they do not eliminate the problem. They direct the leached out salt into the Awash River, causing a rise in the river's salt level and harming downstream dwellers who use the water for irrigation, watering livestock, or home consumption. Soil salinity caused by irrigation has greatly affected crop diversity in the Awash valley. The range of crops that can survive and grow in high salt concentration is very limited, and as such, most irrigation schemes grow only cotton, which can be disastrous for the land as soil nutrients are depleted and not replenished.[15]

An ever increasing number of irrigation schemes and commercial farms also led to increased incidence of water-borne diseases, such as malaria, bilharziasis, cholera, and others. This, coupled with the growing use of herbicides and insecticides, has had adverse effects on the health of both the livestock and the Afar pastoralists.[16]

Indeed, the decline of land available for grazing, the degradation of rangelands, and an increase in pastoralist and animal populations put the continued existence of the Afar in grave danger. Consequently, the Afar have begun to resent the government, commercial farmers, and any outsiders. They have responded by driving their herds into the cotton fields, which are known to mature when the fodder supply in the pastures is low. This has resulted in armed conflicts between the Afar pastoralists and the state farm guards that usually result in the loss of lives on both sides.[17]

The development of the Awash National Park in 1966 is another state development project that has affected the Afar pastoralists. The park covers 830 square kilometers and is in what was formally the Afar wet-season grazing land. The park authorities have been known to confiscate wandering animals and force the pastoralists to pay fines to get their animals back.

Loss of land and restriction of movement have rendered the Afar and their herds more vulnerable to drought. Drought is a recurrent problem faced by the inhabitants of the Awash valley. The Great Famine of 1973–74 is arguably the most severe drought that has ever hit the region. The drought reduced the cattle population by 90 per cent, sheep by 50 per cent, camels by 30 per cent, and goats by 30 percent.[18] It also reduced the Afar population by around 25–30 per cent.[19] Likewise, the drought of 1983–84 had disastrous effects on the Afar and their livestock. However, the problems faced by pastoralists in 1973 and 1983 were not caused by the drought alone. The development of irrigation schemes and consequently the dislocation of the pastoralists from their seasonal patterns to permanent settlements along the river had rendered them very vulnerable to famines. The Afar were forced to travel up to 150 kilometres in search of grazing land water.[20] As the vegetation cover during the periods of reduced rainfall changed from annual grasses to bushes, the pastoralists cut back on grazers (cattle and sheep) and increased the number of browsers (camel and goats).[21]

Perhaps the most adverse consequence of land alienation in the Awash valley has been an increase in inter- and intra-ethnic conflict as pastoral groups fight over the remaining resources. Conflicts have also increased between the Afar and irrigation farm managers and park authorities. The Afar have increasingly become a target for the Issa pastoralists whenever they take their herds to their traditional pasture sites on the Alledeghi Plain. The expropriation of their dry-season grazing lands has also forced the Afar to move into the territory of their neighbours, the Karrayu, Ittu, and Argoba. This has put them in constant conflict with these numerically smaller but well-armed groups. Armed clashes with the Karrayu have become more frequent now than ever before. The Karrayu have also been the victims of land alienation as they have lost access to the Awash River and their traditional dry-season grazing lands in the Metehara, Merti, and Illala plains. Before the establishment of the Metehara sugar plantation on their traditional livelihood space, the Karrayu migrated no more than a 50-kilometre distance from their homes in search of pasture and water. They now have to travel hundreds of kilometres, thus putting them in danger of collision with their Afar and Argoba neighbours who are forced to do the same.[22] Conflicts have also increased between Afar clans. It usually occurs when one clan moves into the territory of another without prior arrangement or permission. Traditionally,

Afar clans allow each other to use resources on their territory, provided that the tenure rights of the host clan are not violated. However, the establishment of large-scale irrigation farms has significantly reduced the amount of land and resources available. This has led to the undermining of traditional inter-clan relations as each clan is compelled to protect its remaining land and resources from neighbouring communities.[23]

There were a few attempts in the early 1970s and 1980s – initially by the Haile Selassie government and later by the military regime – to resettle some of the Afar pastoralists displaced by irrigation schemes. The resettlement schemes were partially conceived as compensation for the concessions of their grazing land. The schemes had three objectives: to turn pastoralists into productive farmers, orient them to a market economy, and transform their lifestyle from nomadic to sedentary. Because the resettlement schemes were mostly influenced by the central government's development objective – the expansion of mechanised irrigation agriculture in the Awash valley – they did not make a significant difference in the lives of those resettled. Several factors accounted for the failure of the resettlement schemes. First, only a minority of the Afar population was included in the project. The amount of land dedicated to the programme was also insufficient to accommodate the large-scale transformation of the Afar from pastoralists to sedentary farmers and ranchers. The Afar received much less land than they had lost to irrigation farm projects. Second, the resettlement schemes concentrated people and livestock into a small area, contributing to overstocking and environmental degradation. Overgrazing resulted in a decrease in total vegetation cover and changes in the floristic structure of the natural vegetation. Third, the economic support promised by the government was insufficient to keep the Afar settlers on their farms. Lack of finances to acquire farm machinery hindered improvement in farm productivity. Fourth, the resettlement brought about the dependence of the local population on food aid and led to a decrease in their standard of living. Fifth, clan leaders and wealthy stockholders benefitted the most from the schemes because they produced cash crops on the enclosed land. For instance, in the Middle Awash, clan leaders such as the Sultan of Aussa distributed communal land to close kin and private investors and made crucial decisions regarding land use without the consensus of the entire clan. The lease of land to commercial farmers by clan leaders further reduced grazing areas. This modern form of a land lease strengthened and enriched the clan leaders and their allies, while most clan members received none of the benefits.[24]

The Afar have been practising sustainable pastoralism for centuries. Seasonal herd mobility, changing herd compositions, and traditional systems were all used to keep the pastoral system ecologically sustainable. However, with changes in land use, economics, politics, demographics, and institutions, Afar pastoralism is increasingly in crisis. Meanwhile, irrigation schemes have failed in achieving the desired results and have instead caused ecological, political, and social problems. Consequently, many Afars are now seeking livelihoods outside the pastoral system.

The few who have remained in their traditional pastoral livelihood have developed innovative means to access grazing resources in the Awash National Park. It involves a 'cut-and-carry system of collecting forage from the park' to feed their cattle outside the park.[25] They haul forage resources with communal and animal labour, such as horse- or donkey-drawn carts. This innovative system has resulted in less confrontation between park authorities and pastoralists, limited the transmission of diseases between wildlife and livestock, and led to increased cooperation between communities to advance other collective economic interests.[26]

The Borana of southern Ethiopia

The Borana pastoralists live in the Borana administrative zone of Oromiya in southern Ethiopia. The zone falls entirely within the Oromiya state and covers about 45,435 square kilometres of land. Slightly over a million people live in the zone, of which 10 per cent are urban residents.[27] The zone is mostly semi-arid and arid, with a few elevated areas being sub-humid. The Genale River forms its eastern border and the Segen River and Chew Bahir form its western border. The Ethiopia-Kenya borderline forms the southern border. A line drawn from the Genale to Segen Rivers forms the northern border, excluding the forested highlands to the north. The Borana have made use of the area's substantial rangelands for several centuries. The main livestock includes cattle, sheep, goats, and camels. Among the few Borana clans who consume camel meat, camel numbers are slowly rising to the level of cattle. Some grow crops such as maize, sorghum, and haricot beans. The bulk of the household income comes from the sale of livestock, milk, and milk products. Over two-thirds of the staple foods are purchased in the market.[28]

The mean annual rainfall ranges from 350 mm in the south and east to a little over 700 mm in the northwest. Rainfall displays significant variations in timing, space, quantity, and distribution, which is characterised by late arrival, early termination, or sometimes total failure. Rainfall is bimodal, with three-fifths of the annual precipitation occurring between March and May and about one-third between September and November. In a typical year, there are 100 to 140 days of plant growth. However, a drought year – when rainfall is below 75 per cent of the average – occurs one out of every five years. Temperatures are usually high year-round. From March to April in the higher areas, the minimum average temperature is above 10 degrees Celsius, while the maximum is around 25 degrees Celsius. During this period, the temperature in the lowlands ranges from 15 to over 35 degrees Celsius. Such high temperatures further reduce the effectiveness of rainfall because moisture is quickly lost to evaporation.[29]

For the most part, the Borana region has an undulating topography. Most of the land lies between 1,000 m and 1,600 m in elevation, and slopes gently downward toward the south and southeast. There are, however, areas that lie above the 1,500-metre mark, such as the eastern towns of Negele and Filtu, the western

Teltele area, and the western-central Yabelo-Mega Plateau. Along the Genale River and the Dawa River, there is a strip of land below 500 metres. Only the upstream of the Dawa and Genale rivers passes through the Borana zone and flows southeast toward Somalia via the Ethiopian Somali regional state. For the most part, these rivers flow through narrow gorges, only opening out onto a floodplain at their lower ends.[30]

Savanna biomes cover most of the region. Vegetation consists of a mixture of perennial grasses, forbs, and woody vegetation. Numerous indigenous species of grasses and woody plants provide high-quality forage. Forage plants tend to increase in nutritional value during the rainy season compared to the dry season. During the dry season, browse plants often retain higher crude protein than grasses. The distribution of plants depends mostly on elevation, temperature, precipitation, and soil types. Surface water is scarce across the Borana plateau. The two major rivers in the region, the Dawa and Genale, are only found in the northeastern corner of the Borana zone. To access water for their livestock during the dry season, the Borana must rely on wells that are extremely deep, which are located mostly in clusters in the west. These wells provide more than four-fifths of the total accessible water during the dry season. Drawing water from wells is labour intensive, and the wells have social and economic significance to the region's pastoralists. The wells also play a key part in determining the seasonal movement of livestock.[31]

The Borana people's primary asset and security are livestock. Each family keeps a herd of mixed livestock, of which cattle are the most important. The herd must be sufficiently large enough to provide enough milk for the family. The herds typically comprise of mostly female animals. The typical ratio of female to male animals is 71:29. The average production unit contains around fifteen cattle, eight of which are milk cows, seven small ruminants such as sheep and goats, and the occasional equine or camel for transport. Milk, meat, and live animals are also often exchanged on the open market for grain, which is considerably more energy-rich. Poorer families often cannot subsist entirely on their animals alone and thus rely on the market, and the rich also sell considerable quantities of live animals for cash. While the herd must be large enough to fulfil these needs, the labour provided by the household may not be sufficient if the herd grows too large, especially when it comes to drawing water from wells. In this case, extra herders can be hired. The Borana often separate their herd into lactating and non-lactating animals; the former are kept near the family home, while the latter are sent to more remote pastures.[32]

Access to the range and pasture is free and collective, although clans, sub-clans, and families dictate the rights and access. Ownership of animals is individual. Water is a limiting factor for livestock production and herd growth, spawning complicated political and familial arrangements. Surface water is communal, with no one having the right to restrict access. However, access to wells and ponds is often related to contributions of labour needed for construction and maintenance. Wells and ponds are often associated with an individual who initiates the project.

The deep wells rely on the cooperative effort of a large group of people and often are beyond the means of operation for a single family. To use the well, a herding family is expected to contribute labour. Each well is connected with a particular clan or lineage and is symbolically owned by an individual. The well's ownership is passed down from father to son. A group of elected officials that form a 'well council' control the operation of the well. The well council not only regulates its daily use and maintenance but also ensures that no physical violence occurs regarding its access. Access to wells is often achieved through kinship, friendship bonds, and exchanges of cattle, along with continued participation in operating the well and the well council's politics.[33]

The movement of the Borana reflects the availability of water and follows a set pattern. Borana herders move their livestock between two sources of water depending on the season. During the wet season, there is abundant surface water across Borana territory. However, during the dry season, water becomes scarcer and herders must move their animals toward deep wells and ponds. The deep wells are unevenly distributed but are clustered in one locality. Movement of Borana herders is also dictated by the communal system, which defines grazing areas based on kinship and familial connections.[34]

The Borana have traditionally used the rangeland sustainably. For many centuries, the Borana have lived in their territorial domain, managing their livestock through customary rules and governing access to pasture and water resources with minimal outside meddling. In recent decades, internal and external factors have compromised their long-established ecologically balanced resource management system. Today, their rangelands have diminished and have been degraded due largely to a host of adverse factors, including government-imposed policies, population growth, encroachment of agriculture, blockage of migration routes, loss of the domain to neighbouring ethnic groups, and frequent drought conditions.[35]

Direct state intervention into the lives of the Borana began in 1960 when the central government introduced development programmes aimed at increasing pastoral productivity. The government built roads, water ponds, and ranches, and also provided veterinary services and market outlets.[36] The military regime in the 1970s introduced compulsory taxes, sale of livestock, state ownership of ranches, resettlement and sedentarisation programmes, and extension services to promote crop cultivation. It also constructed well over a hundred ponds and restored many traditional wells.[37] Under the post-1991 government, rangelands were privatised, pastoralists were encouraged to pursue a sedentary life, and much land was put under crop cultivation.[38]

All these government-induced, top-down changes were pursued outside of the Borana social organisation and traditional water management and production systems. For instance, expanding the water supply proved to be problematic because ponds promoted overgrazing in the area. Since the new ponds were not integrated into the Borana system of resource management, users did not feel bound by the traditional code of conduct in their use. Increased water supply allowed animal numbers to increase, prompting overgrazing, bush encroachment, and extensive erosion. This was most apparent in hilly areas with permanent wells and ponds where pastoralists and their herds tend to congregate during the dry season.[39] As cattle strip the grass cover around wells, woody plants invade, and competition from grasses decreases. Overgrazing of the land has led to decreased milk production. As a result, the Borana have increasingly relied on the market to obtain sufficient food, causing the growing population to become more vulnerable to market fluctuations. The famines of the mid-1970s and mid-1980s in the region were not caused by drought, but rather by the sudden increase in the relative price of grains.[40]

Recurrent drought has been a major problem for the pastoral communities in Borana zone. The impacts of drought include water and pasture shortages, overgrazing, land degradation, livestock deaths, decreased livestock productivity, disease resistance, dislocation of people, and increased conflict over scarce resources.[41] In the past, drought hit the region every 7 to 8 years, but drought frequency has recently risen to every 2 to 3 years. Cattle losses were estimated at between 37 per cent and 42 per cent after the drought of 1983/84.[42] In the 1999/00 drought, an estimated 82 per cent of cattle, 80 per cent of small ruminants, and 60 per cent of camels were lost.[43] The 2006 drought also led to the deaths of up to 60 per cent of all livestock in the area.[44] In response to these disasters, past and present Ethiopian governments have encouraged agricultural livelihoods.

In the Borana rangelands, conflicts have always been linked with inter-ethnic competition over sparse water and grazing resources, especially during drought episodes. The Borana compete with the Guji, Gabbra, Gari, and Somali pastoralists for water and pasture. Up until around a hundred years ago, the Borana controlled much of southern Ethiopia, northern Kenya, and parts of Somalia. The making of colonial boundaries and losses to more powerful neighbouring pastoralist groups significantly reduced their territorial domain. The Borana lost much of their domain once again in the 1990s when the current government redrew administrative boundaries along broad ethnic lines. The Borana lost a vast amount of their traditional grazing lands and vital wells in the process. In a 2003 plebiscite, the Borana lost the entire eastern rangeland – about two-thirds of their rangeland and two of their deepest wells, the Gofa and Lael – to the neighbouring Somali ethnic group and to the newly created Somali regional state in the east. The Borana also lost access to the lower Dawa River. As a result, inter-ethnic clashes between the Borana and Somali communities have become more frequent than ever before, and grazing resources in the border region remain inaccessible.[45] Additionally, the Borana have lost part of their northern territory to the Guji, resulting in a heavier concentration of livestock in the central Borana plateau.[46] All these losses have put the remaining rangelands under enormous pressure. Inter-ethnic conflicts are also a serious obstacle to the Borana livelihood. Conflicts impede customary mobility, destroy lives and assets, and weaken the capacity of pastoral communities to cope with environmental variability and change.[47]

Climate-induced changes and the uncertainties surrounding livestock production are forcing the Borana to shift to agropastoralism. Two decades or so ago, more than one-half of the Borana people pursued pure pastoral livelihoods. Today, it is only 20 percent.[48] Recent years have seen a growing trend of cultivating crops, especially after the droughts of the mid-1980s. Crop cultivation was introduced as a means of augmenting household food requirements and diversifying the means of livelihood. However, most of the cultivators are those who are unable to generate enough income from livestock production or have lost most of their livestock. The proportion of cropland has increased from 3.3 per cent in 1973 to 15.8 per cent of the total land in 2006.[49] Cultivation is confined to the wetter parts of the plateau – often the best grazing land – and consists of small, scattered, half-hectare plots. The principal crops are maize and sorghum, planted during the long rains, and cowpeas planted during the short rains. People grow these crops for both subsistence and household cash earning. For many Borana agropastoralists, crop cultivation has become the second major income earner after livestock. However, despite government and donor-supported provisions of agricultural inputs and extension services, crop cultivation has yet to provide food security. Crop productivity has persistently declined. For example, maize yield declined from 20 quintals per hectares in the 1990s to only seven quintals per hectare at present.[50] This is due largely to the inadequacy and unreliability of rainfall in the region.[51] At any rate, as crop production is increasing at the expense of livestock production, conflict is bound to arise over land use between those who sustain the traditional pastoral mode of livelihood and those who shift to agropastoralism.[52]

Recent years have also seen the state and private ranching on prime grazing land. Private ranches are mostly fenced to prevent free grazing. A survey of 300 pastoral households by Skinner found that three-fifths have lost access to their customary grazing land due to newly created state and private ranches. In the Diid Yabelo area alone, pastoralists and agropastoralists have lost the *Chalalaka* (dry season) and the *Adona* (wet season) grazing areas, along with many water points, including Modi Sooro, Buyii, Ariste, Hardimitu, and Arboji Ponds. In the Surupa area, they have lost the Diid Tuura and Harawet seasonal grazing sites, including the Harbor and Ariste Ponds. At Bulbul, agropastoralists have been affected by the expansion of private enclosures. The expansion of state and private ranches caused all of the losses.[53] There are five large ranches. The Diid Tuura state ranch (5,500 hectares) is state-owned, established for conservation of the Borana breeds and production of heifers for the national breeding programme. The Surupa and Diid Liben ranches (5,525 hectares) belong to a private agro-industrial company (ELFORA) and are used for animal fattening for live export and domestic markets. The Damballa Wachu cooperative ranch (15,000 hectares) is a 170-member group ranch used for animal fattening by members only. The Sarite community ranch (7,750 hectares) is community-managed and used as a fodder reserve for the dry season and can only be accessed by paying fees. The community imposes heavy fines on transgressors or trespassers.[54] Overall, about 60 percent of the Borana households said they have lost a good portion of their prime grazing land to private and state ranches.[55]

As the Borana are slowly squeezed into diminishing rangelands and their customary land use systems deteriorate, land degradation is bound to occur. Today, the Borana rangelands are undergoing rapid land degradation and ecological alterations. Overgrazing, unsuitable agricultural practices, removal of vegetation for firewood and charcoal production, and recurrent droughts have led to increased soil erosion and bush encroachments. Woody plants have replaced perennial grasslands. In Yabelo wereda, the heart of Borana, grasslands have been reduced from 43 per cent in 1973 to just 6 per cent in 2006.[56] In other areas of the Borana zone, *Camphor Africana, Acacia melliphera, A. drepanolobium, A. brevispicia,* and *Lannearivae* species have encroached large areas, replacing perennial grasses and herbaceous species. These species are neither useful for fuel nor provide food for livestock.[57]

The loss of much of their traditional rangelands and water resources and the curtailment of their mobility has exposed the Borana pastoralists to frequent drought and food insecurity. Pastoral households now have only a fraction of the livestock population they had three or four decades ago. Livestock productivity has equally plummeted, owing to diminished access to quality rangelands and water resources as well as recurrent drought. Over 80 per cent of Borana households are now food insecure for about half a year, and nearly one-fifth are food insecure year-round.[58] During these tough times, households have to sell their livestock assets, rely on food aid, or both. Many pastoralists who can no longer make a living from livestock production have dropped out of animal husbandry. Some are increasingly involved in a variety of other income-generating activities to supplement their household incomes. A 2007 study found that over one-third of the Borana households worked in activities other than herding to diversify their incomes.[59] Some of these activities were selling firewood, charcoal, and wooden poles for home building and petty trading of small animals, gum Arabic, and incense. Many pastoralists have also established themselves near trading centres to produce farm products for the market. For instance, poor pastoral households who raise small ruminants and poultry to produce butter, fresh milk, eggs, and meat for the urban market populate peri-urban communities around market towns, such as Agere Maryam, Yabelo, Mega, and Moyale.[60] As more and more Borana pastoralists shift to non-livestock livelihood activities, women's workloads have intensified. Women travel long distances to collect fuelwood, thatching grass, and fodder for sale in the market. Women also work in other activities, such as making charcoal, buying household goods to sell at a higher price, and raising smaller livestock and chickens. Some women work in the urban areas. Often women's income represents the only source of livelihood for the family.[61]

The Somali pastoralists

Much of southeastern Ethiopia is the home of Ethiopian Somali pastoralists and agropastoralists who have an affinity to ethnic Somalis living in Somalia, Djibouti, and northeastern Kenya. Nearly three-fifths of the pastoralists in Ethiopia live in this region. Somali society is based on kinship and residence. Since kinship groups tend to be territorially based, the two are often linked. It is especially true at the level of larger clans as compared to smaller, sub-clans within the larger kinship group.

Membership to clans is determined by patrilineal descent, with the two dominant regional clans being the Dir and the Daarood. Because clan families are so large, they rarely operate as corporate entities, although they do have political significance. It is not unusual for a clan to unite politically for a time against an outside threat and later divide into conflicting segments. Sometimes unrelated clans join in an alliance based on shared principles. Clanship is a potent force in Somali society. Somalis often live among their clansmen in groups of varying size, depending on the situation. Clans often have a territorial base that constantly changes, depending on military strength and the state of inter-clan relationships. Under the period of British administration in the Ogaden clan, boundaries became virtually permanent. Despite these clan territories, movement is usually free across clan boundaries. However, as agriculture becomes more prominent in areas such as Lafeissa, Teferi Ber, Jijiga, and Gursum, these boundaries have become more sensitive. Even with ongoing shifts in the nature of the relationships between clans and territories, it is clear that among smaller, more closely related groups, territorial boundaries become indistinguishable.[62]

The Somali people are traditionally nomadic pastoralists. They depend on their livestock and are constantly on the move in search of water and grazing areas for their animals. Having much livestock in one's herd is a mark of high social status and prosperity. Pastoralists traditionally move across considerable distances with their herds, families, and clan members in search of water and grazing lands. Until recently, there were several lucrative markets for livestock. These were primarily in the Arabian Gulf States. Livestock was transported through Djibouti, Somaliland, or Somalia. However, drought, regulation, and lack of security have largely cut off access to these markets.[63]

Rangelands cover a large portion of the Somali Regional State, which has a total surface area of about 279,252 square kilometres or 26 per cent of the nation's area. The topography of the land follows a gentle slope down to the south and east. Except for shallow depressions of streambeds, the area forms a homogeneous plain. The elevation ranges from 300 m at the farthest southern and eastern borders to up to 1,500 m and along the basins of the Wabi Shebelle and Genale Rivers. The Jijiga Plains in the northeast are exceptions at more than 1,700 m.[64]

Precipitation is a limiting factor in the Somali regional state. An overwhelming proportion of the state receives insufficient rainfall to support cultivation without irrigation. Precipitation ranges between 700 mm of rainfall at higher elevations and below 100 mm at lower elevations in the southeastern parts of the state. Rainfall patterns are bimodal, with a distinct wet and dry season. It is especially true of the lowlands, were the bimodality is most apparent. Rainfall in the highlands is unimodal, so rain patterns become more intermittent at higher elevations.[65] In general, there are five months of rain split into two rainy seasons: from March to May and from October to November. In the lowlands, the dry seasons are from June to September and from December to February. The most rainfall occurs in October, but April and May also receive substantial rainfall. In the highlands, there is also significant rain between June and September, but as in the lowlands, the months of December to February are relatively dry. Temperatures tend to be high year-round, especially in the lowlands. At high altitudes, temperatures range from as low as 10 degrees Celsius to greater than 25 degrees Celsius. At lower altitudes, the temperature is usually between 20 degrees and 35 degrees Celsius.[66]

Surface water is in the form of perennial rivers such as the Wabi Shebelle, Weyib, and Genale, or intermittent rivers like the Fafan, Jerer, Daketa, Erer, Ramis, and other small streams. Most rivers originate in the highlands of Hararghe and Bale and eventually culminate into the Wabi Shebelle and Genale Rivers. Another source of water is groundwater, which people can access via wells. The region accounts for over half of the lowland wells in Ethiopia. The most important aquifers are under the alluvial deposits along the flood plains of the Wabi Shebelle, Fafan, Jerer, Daketa, and Erer Rivers, along with the tributaries of the Wabi Shebelle and the Genale. The sandy soil allows for good groundwater recharge during flooding. The wells dug in these deposits are an important source of water for both people and livestock. However, water quality can be saline due to mineral and salt deposits. In addition to wells, *birkedas*, which are cement-lined cisterns or water tanks, also provide water. During the rainy season, water is directed into the *birkedas*, where it is stored for use during the dry season. This type of water harvesting is especially important in areas such as the lower Ogaden and Elkere.[67] Access to water is dependent on the nature of the source. Surface water is free to all, while wells, cisterns, and man-made ponds are the property of particular individuals or sub-clans. Deep wells tend to be the property of sub-clans, while extended families own shallower wells. Individuals or lineages own the ponds or *hara*. Well-off individuals often own *birkeda*.[68]

The vegetation of the region varies. Dense to scattered shrubs with a grass understory run along the fringe of the highland, accounting for 31.3 per cent of the land area. Scattered open grassland make up 4.1 per cent. Nearly one in four Ethiopian plant species is in the arid southeast. This region has a high diversity of Acacia, and about one-half of the 150–200 *Commiphra* species are here.[69] Dominant grass species of the region are *Chrysopogon aucheri*, *Heteropogon contortus*, *Cenchrus cilaris*, *Cymbopogon* sp., *Dactylactenum agypticum*, *Euphorbia scordifolia*, *Hyperrhenia* sp., *Bothriarchloa radicans*, *Leptotrichium senegalensis*, *Tetrapogon tenellus*, *and Tragus* sp. Common woody species are *Acacia etabica*, *A. commiphora, A. Senegal, A. brevispica, A. depanolobium, A. mellifera, Boswellia* sp., *Balanites aegyptica*, *Dichrostachys cinerea*, *Combretum molle, Delonix elata, Ziziphus spina christi*, and *Grewia villosa*.[70]

Ethiopian Somali pastoralists herd camels, cattle, sheep, and goats with differing proportions of the animals based on ecological and economic consideration. In Jijiga and surrounding areas, sheep are predominant. In the Degahabur zone and the regions bordering Somalia, camels and goats are the dominant livestock. Cattle are more or less evenly distributed across the region but are most conspicuous along the Wabi Shebelle and Fafan Rivers.[71] In 2009, Somali pastoralists had a total of 591,000 cattle, 1.1 million sheep, 1.5 million goats, 255,000 camels, and 118,000 donkeys.[72]

Pastoral or agropastoral practices vary from place to place, depending on the availability of water or rainfall. In the Harshin district (east of Jijiga), for instance, more than four-fifths of the inhabitants practise a traditional pastoral mode of life. Some also grow a few crops, such as maize and sorghum, on 8 per cent of the district's land. Inhabitants of the Kebribeyah district are overwhelmingly agropastoralists. They grow maize, sorghum, and khat as cash crops. They graze their cattle on communal or private fields. Crop residues and hay also serve as cattle feed. The Meiso district is home to the Oromo ethnic group (the Ala, Nole, and Ittu clans) and the Somali ethnic group (the Issa and Hawiya clans). The Oromo clans are predominantly agropastoral, cultivating one to two hectares of land per household. They have private fields ranging from less than one-half hectare to nearly two hectares per household. The Somali clans live entirely off livestock.[73]

Somali pastoralists have traditionally divided rangeland into wet- and dry-season areas. Wet- season grazing areas are open rangelands only used during the rainy season when there is sufficient surface water. Dry-season grazing areas are near permanent or semi-permanent water sources and are often used eight to nine months of the year. Families divide herds into grazing units; milking animals are kept close to the village, while dry animals are much further away. Herders move the dry animals a great distance during the rainy season and closer to permanent water during the dry season. Pastureland in some areas has been suffering from overexploitation, especially near permanent water sources and settlements. Rangelands at higher elevations and in valleys receive higher rainfall and thus have more potential. These areas have especially high concentrations of woody species, which offer good fodder for browsers. To the south and southeast, bushes and thickets increasingly replace this woody vegetation. The grasses in this area are mostly annuals or perennials that flourish shortly after the rains, which will accommodate limited grazing for a few days. An area of high grazing potential in the region is located northeast of Degahabur, around Aware and southeast of Jijiga. This area is one of the few that has escaped overgrazing, mostly due to the absence of available water.[74]

Utilisation of the region's rangeland is not uniform due to inter-tribal conflicts and the uneven distribution of water and forage. The high potential grazing land is mostly in valleys, which are critical for sustaining livestock during the dry season. However, these areas are under increasing threat from cultivation. Jijiga and the surrounding hinterland were once high-value rangeland but have now suffered vegetation loss due to agriculture. In less than two decades of crop cultivation, tons of fertile soil has been lost due to translocation. Similar degradation has occurred in the Jerer valley. On top of this, shallow wells in the area have promoted overgrazing, further degrading the land. Browsers such as goats and camels are deprived of fodder due to clearing of bushes and shrubs for fire wood and charcoal.[75]

Though the nomadic pastoral mode of life is still a major livelihood in the Somali regional state, more and more pastoralists have been moving toward agropastoral activities in the past three decades. This change is seen in the increase of range enclosures, fencing, and permanent settlements that rely on cultivating sorghum, maize, and khat. The development of *birkeda* facilitated the growth of permanent settlements and the development of agriculture caused livestock movements to become more restricted.[76] About 400,000 hectares of land are under cultivation in the Somali regional state. Recent years have also seen the expansion of privately operated large- and small-scale irrigation farming schemes in the state; about 28,000 hectares of pastureland has been put to use. In the Kebribeyah district, almost one-third of the land is being cultivated.[77] Many nomadic clans, such as the Gadabursi, Yabarre, Gerri-Jarso, Bartire, and the Abaskul in the Jerer valley, have also turned to farming in high rainfall areas of Jijiga, Teferi Ber, and Gursum. This has been followed by a shift from communally to privately operated land.[78]

Several factors drive the transition to farming. Frequent droughts in the last three or more decades have made pastoralism a less viable livelihood. There were major droughts in 1984–1985, 1994, and 1999–2000 in which pastoralists lost most of their animals due to lack of food and water; they lost 70–90 per cent of their cattle during the last drought alone. Shrinking grazing land also made it more difficult for pastoralists to pursue the traditional, mobility-based mode of life. As a result, many pastoralists have moved from pastoral activities toward agropastoral and farming activities. Second, many Ethiopian Somalis acquired sedentary life and farming skills while living in refugee settlements in Somalia after the 1977–78 war between Ethiopia and Somalia. Upon their return to Ethiopia in the 1990s, many of the former refugees opted to pursue a sedentary life of cultivating crops with little or no animals to keep. Third, the Arabian Gulf State had long been a major market for livestock. However, in the 1990s, outbreak of Rift Valley fever prompted many Gulf States, including Saudi Arabia, to impose a ban on livestock import from the Horn of Africa. The ban continued for several years and severely affected Ethiopian Somali pastoralists, forcing many of them to take up agricultural activities.[79]

The shift toward sedentary agropastoral and farming livelihoods has led to land privatisation and enclosure. Land enclosures began in the 1980s and picked up momentum since the 1990s. Enclosures, in return, have resulted in increased discord between those accustomed to communal ownership of land, such as pastoralists, and those who prefer private ownership of land, such as agropastoralists and agriculturalists. Pastoralists' wet-season grazing grounds are increasingly being encroached on by agropastoralists and agriculturalists. The transition toward agropastoralism has intensified the conflict over valuable land with easy access to water. This land is usually found on or near a riverbank. Demand for such land has doubled in the last two decades. When conditions drive desperate pastoralists to seek water and pasture beyond their traditional territorial areas, they come into conflict with other clans and sub-clans, as well as with agriculturalists. Pastoralists increasingly must bring their animals to water by passing through agricultural land that often belongs to another clan. Conflicts spark when pastoralists' animals trample over crops in the field belonging to another clan.[80]

The region has thus been affected by an increase in privatised water points, grazing enclosures, permanent settlements, and crop cultivation in areas previously reserved for dry-season grazing. Armed conflict between the government and various liberation front groups in the region has also displaced thousands of people, made large areas of grazing resources permanently inaccessible, and disrupted established trading networks.[81] Also, the civil war in Somalia has resulted in an influx of tens of thousands of refugees into the region, further intensifying the pressure over pasture and water resources. The federal government has recently put in place a resettlement plan to relocate scattered households in villages in less arid areas. The government has already resettled about 150,000 agropastoralists living in the harsh environments in the eastern Somali regional state to more fertile areas along the Wabi Shebelle River.[82]

Agropastoral communities

The agropastoral economy is more mixed than that of the pastoral; no one activity can be relied on exclusively to meet subsistence needs. One of the major economic activities of agropastoralists is crop cultivation. Farming is practised in areas that allow access to adequate water, such as along riverbeds or lakes, or in high mountain ranges or plateaus that receive more rainfall. Small plots are typically cultivated using simple tools, such as digging sticks, though some use draft animals for ploughing.[83]

The agro pastoral systems are found mostly in southwestern Ethiopia, including the southern Rift Valley, Omo flood plains, the lower Omo in SNNP, and the Baro River plain in Gambela state. Both crop cultivation and livestock raising are practised in this farming system. Most agro-pastoralist groups live along the Baro, Omo, Kibish, and Segen Rivers and on the Hamar mountain ranges and the Nyalibong Hills. Altitude ranges from below 300 m at the lowest end of the Omo River and around Lake Turkana to over 1,500 m in the Hamar ranges. The Omo River collects fertile top soil from the western highlands and deposits it on a flat plain. The variety of crops grown in agropastoral communities is very limited, compared to both the cereal and enset-root farming systems. The dominate crop is maize, accounting for more than half the cultivated land. Crops such as sorghum, pulses, tobacco, and pumpkins are also produced, but in lesser quantities. Before and after the flood, clearing and burning prepare the land. Planting begins after the main flood in October and November. Sometimes two crops of sorghum can be obtained. However, the river does not always flood at the same time or to the same degree, making farming difficult. Farming tools are primitive and not very efficient; planting is done with a single sharp stick. The case is different in the Hamar uplands, and the Nyalibong and Mursi Hills, where the Omo River does not flood the land and farmers must rely on rainfall. In some parts of Hamar, the pastoralists and farmers in the cereal and enset-root farming systems prepare their land using oxen.[84]

Rainfall tends to be low, from around 800 mm in the north to less than 200 mm in the south. There are two rainfall patterns: with the northwest being mainly unimodal while the southeast is distinctly bimodal. In the northwest, there are eight rainy months from March to October with rainfall fairly evenly distributed. January is the driest month. In the south and east, there are eight rainy months separated into two wet seasons that run from February to July and September to October. The most rain falls during the months of February to May and September to October. Temperatures range from 15 degrees to 20 degrees Celsius in the highlands and from 25 degrees to 35 degrees Celsius in the lowlands. In lowland areas, temperatures tend to be higher during the rainy seasons.[85]

Several agropastoralists groups live in the northernmost part of the Omo gorge and the Nyalibong and Mursi Hills. The Mursi cultivate the riverbanks, berms, and islands that are seasonally inundated. In September, the land is prepared and planted with sorghum, maize, beans, and cowpeas. Harvest is in December and January. In February, the dense bush thickets on the alluvial flats along the tributaries of the Omo are burned and then planted with sorghum and cowpeas during the March and April rains. The crops are harvested in June and July. The Mursi herd their cattle in wooded grasslands as far east as the Nyalibong Hills, where the tsetse fly is absent.[86] To the south of the Mursi, the Bume cultivate the banks of the Omo and the Kibish. A small degree of rain-fed agriculture is also practised if the situation permits. The Geleb farm further south on the banks of the Omo River and in the delta near Lake Turkana. After the rainy season, the growing season begins along the river as soon as the floodwaters have receded. Usually there is enough water for only one harvest, although sometimes a second crop can be cultivated.

The Arbore grow crops around Chew Bahir. The Weito and Sagan Rivers run down the natural slope toward the basins along the Hamar Range. The Hamar, Arbore, and Tsmai used to take advantage of this by redirecting the water from the rivers into irrigated fields. However, the waters of these rivers have receded to the point that now the only water collected in the basins is from rainfall.[87]

The Bode herd their livestock around the margin of the Dime highland. During the dry months from December to February, the cattle are brought to the Omo River. Between May and June, foodstuff is scarce, and some subsistence hunting must occur. Between Narok and the delta, the Omo River has no flood plain. The river here is straight and incised; crops can only grow on narrow and intermittent berms. There are many sites that could be used for small plots in the delta, but the time available for cultivation is limited because the Omo River tends to back up near the lake due to the annual floods.[88]

The Dassenetch grow sorghum, maize, tobacco, beans, and several types of gourds. They prepare the land in August and September before the flood peaks. After the main floods recede, which occurs in October along the river and November to December in the delta, planting takes place. Crops are harvested in December and January on the berms and January and February on the delta. Above Narok and in the delta are oxbow channels and back swamps. Here one and sometimes two crops of sorghum can be grown.[89]

The Nyangatom live along the Kibish River and practise a form of flood retreat cultivation. The narrowness of the berms and flats along the river limits cultivation.

People produce two crops per year: one in March to June and another in September to January. Cattle are grazed about 10 km to 15 km from the river from May to July, moved out as far as 50 km from September to November, and kept close to the river during the dry months of December to February. Cultivation and livestock are supplemented by gathering wild plants.

The Hamar are found in the Hamar Mountains east of the Omo. They rely primarily on crop cultivation but raise some small stock and cattle. The herds graze in the highlands during the wet season and remain in the lowlands during the dry season, where they can be taken to the rivers.[90] They also cultivate the banks of the Koke River to a limited degree. Grass cover was previously ample in the foothills of the Hamar range, making the area suitable for cattle. At present, however, bush encroachment has caused many herders to shift to goats.

The Banna also inhabit the foothills of the Hamar range. They grow varieties of sorghum with different tastes and growth rates. Usually, several varieties are planted to reduce the risk of crop failure. Other crops grown in this area include beans, squash, sweet potatoes, tobacco, red pepper, and cotton. They plant the crops during March rains and harvest in May. In years of extended rainfall, a second crop of sorghum can be obtained in June and July. Sometimes there is enough rainfall during the short rains in September and October to produce another crop. The Banna subsist mainly on their agricultural production, but their society is more oriented toward livestock. Cattle herds are divided into two groups: lactating cows and calves are kept near the homestead, and the non-lactating herd are moved from pasture to pasture based on the availability of forage and water and to avoid tsetse flies.[91]

Agropastoralists also inhabit the northern end of the Gambela plains. Except for a narrow strip of steep and rugged land between 700 m and 1,000 metres of altitude, the land follows a gentle slope, hitting a low point of about 400 metres at the Ethiopian-Sudanese border. Several rivers cross the plain from east to west, such as the Baro, Alwero, Gilo, and Akobo. These rivers often become waterlogged during the wet season.

This region is distinct in that it receives a high amount of rainfall and is, thus, endowed with surface water. The lowlands typically receive around 800 mm and the eastern highlands around 1,200 mm. Rainfall is unimodal, with the rainy season running from April to October. The most precipitation falls between June and September. In the lowlands, temperatures range from 20 to 35 degrees Celsius; in the highlands, the temperature varies from slightly below 20 degrees to over 30 degrees Celsius. Three to four major rivers run through the area and during the rainy season, the parts of the plain less than 425 metres above sea level become flooded for several months. During the flooding, pastoralists migrate to the highlands. Livestock production is not as high here as compared to other regions due to an infestation of the tsetse fly and trypanosomiasis. As a result, many ethnic groups are agropastoralists rather than pastoralists.[92]

The Nuer of the Gambela state cultivate both along the banks of the Baro River and in the uplands. The Nuer base their movements on the flooding of the river. For four months, beginning in May, the Baro floods. During the flood period, the Nuer cultivate the Lare highlands, planting maize or sorghum between May to June and harvesting from August to October. Once the floodwaters have receded, they return to the river banks to grow crops and to graze their livestock on communally held pastures.[93]

Another significant activity in these agropastoralists regions is fishing. The Nuer, Geleb, and Arbore all take advantage of this resource. They use both hooks and gill nets. The Nuer use cotton, which are suspended between two canoes and allowed to float downstream. Fishing usually takes place at night during the dry seasons from November to April. Fish are preserved by drying and salting. Another method is spear fishing, which is utilised when the water is reasonably clear. The Nuer fish on the Baro River, whereas the Arbore fish on Chew Bahir, and the Geleb fish on the Omo River. The species fished include Nile Perch, Flounder, Cat Fish, and Snake Fish, as well as crocodiles and tortoises. Some agropastoralists also engage in hunting and gathering.[94]

In all these agropastoral societies, the local culture places great importance in livestock though its management suffers from the fact that the genetic stock has adapted to survival rather than production. As such, productivity is low even by the standards of the area, as is the demand for products, such as milk, which further constrains development. There are also other significant constraints to live-stock development, including low-quality forage, shortages of both forage and water during the dry season, and disease (mostly trypanosomiasis), especially for cattle. Because of the disease, most of the livestock are sheep, goats, and donkeys. Goats browse, which makes them able to handle the sparseness of the forage. Though there are few camels in the area, their status as the browsing animal makes the area likely to serve as good pasture for the camel.[95]

The main constraints to the agropastoral economy have to do with the type of cultivation in which each group engages. In systems that rely on flood retreat crop-ping, such as those along the Omo and Kibish rivers, there is a limited amount of suitable cultivation sites. The degree and timing of floods also limit the quantity of crops that can be produced. Population growth has caused the subdivision of the family plots that are already too small. This has aggravated severe seasonal food deficits. In regard to pastoralism, the success of livestock raising depends on the availability of grazing lands that have adequate forage and are free from tsetse flies. Livestock production is hindered by the fact that large swaths of rangeland are incorporated in the Mago National Park and Tama Wildlife Reserve. The pres-ence of so many different ethnic groups within a small area has led to tribal ten-sions and constrained livestock production.[96]

For agropastoralists engaged in shifting cultivation, such as those in the Hamar Uplands, rapid population growth has put severe pressure on the land. For example, population growth has led to illegal encroachments into state forests. Forests are used for grazing or coffee production, as has happened in the Angetu area, near Delomena. In the Hamar Uplands, extensive tracts of *combretum* woodland have been cleared and used for cultivation. It appears that little of the land is allowed to rest and revert to woodland, as was the traditional practice.[97]

The most serious threat to the livelihoods and identity of agropastoralists in the Omo River Valley comes from the ongoing state-sponsored water development projects in the valley (for more discussion on this subject, see Chapter 5). A series of dams built on the Omo River and its major tributaries have displaced thousands of indigenous communities whose well-being and livelihoods depend on the integrity of the resources along the river. The more than 200,000 indigenous people who live along the banks of the river have also lost 245,000 hectares of land to a government-run sugar plantation. The Lower Omo Valley is designated a UNESCO world heritage zone because it is among 'the most culturally distinctive and biologically diverse valleys in the world.'[98]

Conclusion

Pastoralism is the main source of livelihood for the vast majority of the people living in southern Oromiya, Somali, and Afar regional states. Small communities in southern SNNP, Gambela, and Benshangul-Gumuz regional states also subsist on pastoral and agropastoral forms of livelihood. Ethiopian pastoralists face numerous constraints in pursuit of their livelihood.[99] Over the last half a century, pastoralists' rangelands have been shrinking as more and more land is converted into irrigation and dry farming. The 1994 Constitution states: 'Ethiopian pastoralists have the right to free land for grazing and cultivation as well as the right not to be displaced from their lands.' In reality, however, the traditional rangeland of pastoralists has been reduced by the expansion of sedentary agriculture, commercial agricultural projects, national parks and reserves, and water development projects. The rangelands are in many cases bordered at their highest altitude by farmland. The upper limits of grazing land are relatively rich in escarpments receiving between 400 mm and 800 mm of rain. These favourable conditions invite agriculturalists to encroach on pastoralists' land. Such has been the case in the regions of the Afar, Somali, and Borana, where pastoralists have been under pressure from highland farmers and government-initiated development schemes.

As pastoralists are pushed away from their traditional pastures, they are forced to induce overgrazing on the remaining land available for pasture. Traditionally, pastoralists have coped with drought by moving to areas with more forage and water. However, this movement is now constrained as settlers, whose numbers have grown due to population growth, have occupied many former grazing grounds. The degradation of the country's pastoral or rangeland is not only detrimental to the livelihood of several million people but also exacerbate the adverse impact of climate change since rangelands capture considerable volumes of carbon in shrubs, bushes, and grasses.

Overgrazing has resulted in the emergence of some invasive alien plant species, which are posing a major threat to the livelihoods of pastoralists. *Prosopis juliflora*, which was purposely introduced to promote agroforestry, is wreaking havoc in the Awash National Park, Upper Awash Valley, and Eastern Hararghe. The plant is 'destroying natural pasture, displacing native trees, forming impenetrable thickets, and reducing the grazing potential.'[100] *Parthenium hysterophorus*, which was introduced unintentionally through relief aid consignment, is wildly spreading in the Afar and Somali rangelands and has also reached the Borana pastoral lowlands.[101]

The diminution of pastoral resources has resulted in frequent and destructive inter-clan conflicts. Inter-tribal conflicts have far-reaching consequences. Besides the loss of property and life, these conflicts can cause the inefficient use of resources. For example, a conflict between the Afar and Issa Somali has prevented the use of 75,000 hectares of grazing land on the Alledeghi plain for fear of attack. Other conflicts between the Somali, Borana, and Omo create a situation in which resources are made unavailable altogether or are made available to one group at the expense of another.[102]

Pastoral areas lack transportation infrastructure. All areas have at least one major all-weather road, and the Afar and Borana regions do have some paved roads. The Borana region is reasonably accessible due to the work of the Southern Rangeland Project, but the lack of roads in other regions limits access. Many pastoral areas are reachable in the dry season only by four-wheel drive vehicles. Pastoralists are also disadvantaged by the distribution of markets, which are often at the very edge of their territory. Pastoralists must often walk for days before reaching market centres and bear the economic cost of any losses incurred along the way.

Notes

1 Eyasu Elias. 2008. *Pastoralists in Southern Ethiopia: Dispossession, Access to Resources and Dialogue with Policy Makers*. Drylands Coordinating Group Report No. 53, p. 1. Available online at www.drylands-group.org/publications/pastoralists-in-southern-ethiopia (accessed 9 July 2016).
2 Richard Hogg (ed.). 1997. *Pastoralists, Ethnicity and the State in Ethiopia*, London: Haan Publishing, pp. 11–3.
3 Eyasu Elias, p. 1.
4 Water Resources Development Authority. 1989. *Amibara Irrigation Project II Pastoralists and Forestry Development Studies*. Final Report, p. 2.3.
5 Red Cross Society. 1987. *Pastoralist Rehabilitation and Development: Irrigation Projects in the Gewane and Assaita Areas*, Addis Ababa, p. 2.
6 Kassa Negussie Getachew. 2001. *Among the Pastoral Afar in Ethiopia: Tradition, Continuity and Socio-economic Change*, Utrecht, The Netherlands, p. 37.
7 Getachew, pp. 37–9.
8 Ibid., pp. 46–7.
9 Water Resources Development Authority, p. 3.3.
10 Helmut Kloos. 1982. Development, Drought, and Famine in the Awash Valley of Ethiopia. *African Studies Review* 25, 4: 21–48; pp. 5–7.
11 Lars Bondestam. 1974. People and capitalism in the northwestern lowlands of Ethiopia. *Journal of Modern African Studies* 12, 423–39.
12 Ali Said. 1997. Resource use conflict in the middle Awash Valley of Ethiopia: the crisis of Afar pastoralism. In: *Pastoralists, Ethnicity and the State in Ethiopia*, Richard Hogg (ed.), London, Haan Publishing, pp. 105–24; p. 124; Getachew, pp. 90–1; Andrew Ridgewell, Getachew Mamo, and Fiona Flintan (eds.). 2007. *Gender & Pastoralism,*

Volume 1: Rangeland & Resource Management in Ethiopia, Addis Ababa, p. 49. Available online at www.sahel.org.uk/pdf/Gender%20&%20Pastoralism%20Vol%20 1%20-%20ebook.pdf. (accessed 12 November 2010). The Awash River basin holds about 71 percent of the country's irrigated land (Irit Eguavoen). 2009. *The Acquisition of Water Storage Facilities in the Abbay River Basin, Ethiopia.* Working Paper Series 38, Center for Development Research (ZEF), University of Bonn, p. 5.

13 Peter Koehn. 1979. Ethiopia: famine, food production, and changes in the legal order. *African Studies Review* 22, 1: 51–71; p. 55.

14 Sir MacDonald and Partners. 1987. *Kesem Irrigation Project Feasibility Study.* UNEP/ FAO Report to WRDA, Volume 2, Annexes A–D.

15 Ali Said. 1994. *Pastoralism and State Policies in Mid-Awash Valley: The Case of the Afar, Ethiopia, African Arid Lands.* Working Paper Series No. 1.

16 Helmut Kloos and A. Lemma. 1977. Bilharziasis in the Awash Valley: epidemiological studies in the Nura Era, Abadir, Melka Sedi and Amibara. *Ethiopian Medical Journal* 15: 166–8.

17 Said, 1994.

18 J. Helland. 1980. *An Analysis of Afar Pastoralism in the Northeastern Rangeland of Ethiopia*, African Savannah Studies, Occasional paper No. 20, pp. 79–134.

19 Bondestam, pp. 423–39.

20 Desta Asfaw. 1993. *Large-Scale Agricultural Development and Survival Issues among Pastoralists in the Awash Valley.* Conference on Pastoralism in Ethiopia, 4–6 February, Addis Ababa.

21 Sir MacDonald and Partners, Annexes A–D.

22 Eyasu Elias and Feyera Abdi. 2010. *Putting Pastoralists on the Policy Agency: Land Alienation in Southern Ethiopia, International Institute for Environment and Development (IIED).* Gate Keeper Series # 145, July, p. 10. Available online at www .iied.org/pubs/pdfs/14599IIED.pdf (accessed 16 November 2010).

23 Getachew, p. 97.

24 Getachew, p. 100; John W. Haberson. 1978. Territorial and development politics in the Horn of Africa: the Afar of the Awash Valley. *African Affairs* 77, 309: 479–98; pp. 490–4.

25 Yohannes Gebre Michael, Saidou Magagi, Wolfgang Bayer, and Ann Waters-Bayer. 2011. *More than Climate Change: Pressures Leading to Innovation by Pastoralists in Ethiopia and Niger.* Paper presented at the International Conference on the Future of Pastoralism, 21–23 March, Organised by the Future Agricultures Consortium at the Institute of Development Studies, University of Sussex and Feinstein International Center of Tufts University. Available online at www.prolinnova.net/sites/default/files/ documents/news/gebremichael_et_al.pdf (accessed 17 August 2011).

26 Ibid.

27 CSA. 2010. *Statistical Abstract 2009*, p. 36.

28 Save the Children (UK). 2009. *Climate-Related Vulnerability and Adaptive-Capacity in Ethiopia's Borana and Somali Communities.* Final assessment report, p. 12. Available online at www.iisd.org/pdf/2010/climate_ethiopia_communities.pdf (accessed 4 November 2010).

29 D. L. Coppock. 1994. The Borana Plateau of Southern Ethiopia: Synthesis of Pastoral Research, Development and Change, 1980–91, Addis Ababa: International Livestock Research Institute (ILRI); Mekbib Mammo, Girma Berhanu, and Taddese Yiberta. 1984. *The Nomadic Areas of Ethiopia Study Report.* Part 2 – The Physical Resources, Addis Ababa: United Nations Development Program, Eth/81/001, p. 71; Woody Biomass Inventory and Strategic Planning Project (WBISPP). 1995. *Socio-Cultural and Economic Aspects of Crop, Livestock and Tree Production*, Annex 3, WBISPP, 30 November, p. 94.

30 Mekbib Mammo et al., pp. 71–2.

31 WBISPP, p. 94.

32 Johan Helland. 1997. Development interventions and pastoral dynamics in southern Ethiopia. In: *Pastoralists, Ethnicity and the State in Ethiopia*, Richard Hogg (ed.), London: Haan Publishing, pp. 62–4.

33 Ibid., pp. 67–8.

34 Girma Berhanu and Taddese Yiberta. 1984. *The Nomadic Areas of Ethiopia Study Report: Part 3 – The Socio-economic Aspects*, United Nations Development Program, Addis Ababa, pp. 11–2.

35 Djihan Skinner. 2010. *Rangeland Management for Improved Pastoralist Livelihoods: The Borana of Southern Ethiopia*. MA thesis in Development and Emergency Practice, Oxford Brookes University, April, p. 10. Available online at www.brookes.ac.uk/research/cendep/dissertations/DjihanSkinner.pdf (accessed 9 July 2016).

36 Sabine Homann. 2005. *Indigenous Knowledge of Borana Pastoralists in Batural Resource Management: A Case Study from Southern Ethiopia*, Cuvillier Verlag, Gottingen, p. 54.

37 G. Oba. 1988. *Borana Lowland Pastoral Development Program (BLPDP/GTZ)*, Negelle, Ethiopia. Cited in Skinner, p. 31.

38 Homann, pp. 53–4.

39 WBISPP, p. 96.

40 Helland, p. 72.

41 Save the Children (UK). 2009. *Climate-Related Vulnerability and Adaptive-Capacity in Ethiopia's Borana and Somali Communities*, Final assessment report, p. 18. Available online at www.iisd.org/pdf/2010/climate_ethiopia_communities.pdf (accessed 4 November 2010).

42 Solomon Desta. 1999. *Diversification of Livestock Assets for Risk Management in the Borana Pastoral System of Southern Ethiopia*. PhD dissertation, Utah State University. Cited in Homann, p. 44.

43 Oromiya Disaster Prevention and Preparedness Bureau (ODPPB). 2000. *Loss Assessment Report in Lowland Weredas of Borana Zone*, Addis Ababa: ODPPB. Cited in Homann, p. 44.

44 Save the Children (UK), p. 18.

45 Skinner, 34; Sabine Homann, p. 53.

46 Solomon Desta, cited in Solomon Desta and D. Layne Coppock. 2004. Pastoralism under Pressure: Tracking System Change in Southern Ethiopia, *Human Ecology* 32, 4(August): 465–86, p. 477. Available online at www.springerlink.com/content/p2146461316643t1/ (accessed 9 November 2010).

47 Save the Children (UK), p. 18.

48 Aklilu Amsalu and Alebachew Adem. 2009. *Assessment of Climate Change-Induced Hazards, Impacts and Responses in Southern Lowlands of Ethiopia*. Research Report No. 4, Addis Ababa: Forum for Social Studies, p. 55.

49 Skinner, p. 36.

50 Kejela Gemessa, Bezabih Emana, and Waktole Tiki. 2006. *Livelihood Diversification in Borana Pastoral Communities of Ethiopia – Prospects and Challenges*, Addis Ababa. Available online at www.saga.cornell.edu/saga/ilri0606/brief11.pdf (accessed 9 July 2016).

51 D. Layne Coppock and Solomon Desta, pp. 465–86; Cited in Djihan Skinner, p. 31.

52 Andrew Ridgewell, Getachew Mamo, Fiona Flintan (eds.) p. 31.

53 Eyasu Elias and Feyera Abdi. 2010. Putting pastoralists on the policy agenda: land alienation in Southern Ethiopia, International Institute for Environment and Development (IIED), Gate Keeper Series #145, July, p. 8. Available online at www.iied.org/pubs/pdfs/14599IIED.pdf (accessed 16 November 2010).

54 Ibid., p. 11. Skinner, p. 36.

55 Eyasu Elias, p. 11.

56 Skinner, p. 38.

57 Save the Children, p. 47.
58 Eyasu Elias, p. 19.
59 Ibid., p. 21.
60 WBISPP, p. 98.
61 Ridgewell et al., p. 22; CSA, *2010. Livestock and Livestock Characteristics*, Agricultural Sample Survey 2009, Statistical Bulletin 468, Addis Ababa: CSA, p. 39.
62 Richard Hogg. 1997. Changing land use and resource conflict among Somali pastoralists in the Haud of south-east Ethiopia. In: *Pastoralists, Ethnicity and the State in Ethiopia*, Richard Hogg (ed.), London: HAAN Publishing: 105–22; pp. 110–2.
63 Canadian Hunger Foundation International (CHF). 2006. Grassroots conflict assessment of the Somali region, Ethiopia, August. Available online at www.globalcommunities .org/publications/2006-somalia-conflict-assessment.pdf (accessed 9 July 2016).
64 Mekbib Mammo et al., pp. 58–9.
65 Ibid., p. 59.
66 Ibid., pp. 59–60.
67 Ibid., p. 60–1 and 63–6.
68 Hogg, p. 114.
69 Environmental Protection Agency (Ethiopia). 1997. *A Draft Proposal for the Establishment of National Desertification Fund of Ethiopia*, Addis Ababa: EPA, p. 3.
70 Ibid., p. 67.
71 Girma Berhanu and Taddese Yiberta. 1984. *The Nomadic Areas of Ethiopia Study Report*. Part 3 – The Socio-economic Aspects, Addis Ababa: United Nations Development Program, pp. 10–22.
72 CSA. 2010. *Livestock and Livestock Characteristics*: *Agricultural Sample Survey 2009*, Statistical Bulletin 468, Addis Ababa: CSA, p. 39.
73 Fekadu Beyene. 2009. Exploring incentives for rangeland enclosures among pastoral and agro-pastoral households in eastern Ethiopia. *Global Environmental Change* 19, 4(October), pp. 494–502.
74 Ibid., pp. 67–9.
75 Ibid., pp. 69–70.
76 Hogg, p. 117.
77 Fekadu Beyene, pp. 494–502.
78 United Nations Development Program Emergencies Unit for Ethiopia (UNDP – EUE). 1988. *South East Rangelands Project (SERP)*, Addis Ababa: UNDP.
79 CHF International, p. 15. Available online at www.globalcommunities.org/publications/ 2006-somalia-conflict-assessment.pdf (accessed 9 July 2016).
80 Ibid.
81 John D. Unruh. 2006. Changing conflict resolution institutions in the Ethiopian pastoral commons: the role of armed confrontation in rule-making. *GeoJournal* 64(Spring): 225–37; p. 230.
82 William Davison. 2010. Ethiopia relocates 150,000 people in eastern Somali region in five months. *Bloomberg News*, November 29, 2010. Available online at www.bloomberg .com/news/2010-11-29/ethiopia-relocates-150-000-people-in-eastern-somali-region-in-five-months.html (accessed 4 December 2010).
83 Girma Berhanu and Taddese Yiberta, p. 51.
84 WBISPP, pp. 73–5.
85 Mekbib Mammo et al., pp. 82–3.
86 WBISPP, p. 102.
87 Girma Berhanu and Taddese Yiberta, pp. 52–3.
88 WBISPP, p. 102.
89 Ibid.
90 Ibid.
91 Ibid., p. 103.

92 Mekbib Mammo et al., pp. 87–8.
93 Girma Berhanu and Taddese Yiberta, p. 93.
94 Ibid., p. 53.
95 Ministry of Water Resources. 1960. Omo-*Gibe River Basin Integrated Development Master Plan Study*. Final Report, Volume III, Development Zones and Areas.
96 WBISPP, p. 104.
97 Ibid.
98 Ben Rawlence. 2012. Why are Ethiopian eyes on Brussels? EuropeanVoice.com. Available online at www.europeanvoice.com/article/2012/july/why-are-ethiopian-eyes-on-brussels-/74898.aspx (accessed 24 July 2012).
99 For detailed discussion of pastoral constraints, see Girma Berhanu and Taddese Yiberta, pp. 92–8; WBISPP, pp. 93–8; Richard Hogg, Changing land use and resource conflict among Somali pastoralists in the Haud of southeast Ethiopia, *Pastoralists, Ethnicity and the State in Ethiopia*, Richard Hogg (ed.), London: HAAN Publishing, 1997.
100 Jonathan McKee. 2007. *Ethiopia: Country Environmental Profile*, p. 43.
101 Ibid.
102 Girma Berhanu and Taddese Yiberta, p. 95.

4 Forests and woodlands

Forests and woodlands are vital to Ethiopia's ecological well-being as well as to its citizens' livelihoods. They provide materials for construction, fuel, animal feed, medicine, and industrial products. Forests also serve as natural sinks that help keep heat-trapping carbon dioxide out of the atmosphere and provide habitat for the country's numerous species of birds, other animals, and plants. Ethiopian flora is very diverse. This diversity can be explained by prevailing environmental conditions, including rainfall, temperature, soil, and topography. Unfortunately, however, considerable damages have been done to the country's forest and woodland resources, especially to those found in the central and northern highlands of the country. This chapter deals with the extent and distribution of forests, the degree and underlying causes of forest and woodland degradation, the fate of the remaining forest and woodland resources, and challenges of protecting, managing, and enhancing forest resources.

Degradation of forests and woodlands

The highlands, with their temperate climate and fertile soils, have historically contained most of the Ethiopian population. As a result, it is these areas, especially those regions with the longest human habitation, which have experienced the most land degradation. No other highland ecosystem has undergone ecological destruction as extensively as the northern Ethiopian highlands – the oldest settled region in the country, covering much of today's Amhara and Tigray regional states. It was here that the oxen-plough agricultural system began over 2,000 years ago. The earliest cities and towns, and eventually the first Ethiopian state, emerged in this area. Axum (Tigray), Gondar, and Lasta (Amhara) were seats of great kingdoms that left behind impressive and lasting architectural works. It was also from these highland states that ancient Ethiopian kingdoms were known to have exported hardwood tree species, such as *Ficus*, *Chlorophora*, and *Morus*, to the Middle East and North Africa for use in shipbuilding.[1] Regional kingdoms frequently fought one another over territory throughout the northern highland's history. In the process, villages, farms, and forests were destroyed. A military tactic often employed was the destruction of forests to eliminate hiding places for political opponents. The peasantry and environment endured the brunt of these conflicts. Farmland had to be cultivated, and forests cleared to supply the various armies. When they had exhausted the resources in one place, many rulers simply took their armies and moved on. In many cases, the lack of a permanent royal residence can be partly explained by the lack of firewood in the area.[2]

The historical records of the nineteenth century indicate that large expanses of territory surrounding most settlements in highland Ethiopia had been deforested. Just like wildlife began to disappear with the introduction of firearms, trees, which once probably covered the countryside, mostly disappeared as they were cut down or burned. Some have questioned the extent, causes, and rate of deforestation and the belief that 40 per cent of the highlands were forests about a century ago.[3] No one will ever know exactly how much of the land had forest cover, but remnants of ancient trees in enclosed church compounds, small patches of woodlands of the same species and types provide evidence that the forest was previously much more widespread. Darbyshire and Lamb's analysis of charcoal and pollen in sediment cores from two lakes in the highlands of northern Ethiopia also provide credible evidence that human-induced natural vegetation changes have occurred during the last 3,000-year period.[4] Their study shows that *Podocarpus-Juniperus* forests covered large areas of the region but increased human livelihood activities cleared these forests, and secondary vegetation of *Dodonaea* scrub and grassland replaced these forests for the last 1,800 years. Grasslands dominated from about 1200 to 1400, probably due to grazing intensification or aggravated by drought conditions. '*Juniperus* forest, with *Olea* and *Celtis*, then expanded from 1400 to 1700, possibly because of drought-induced depopulation followed by increased rainfall.'[5] Their study concludes that increased human settlements and activities deforested and eroded the soil again in the last three centuries.

Centuries of overexploitation have rendered the northern highlands an ecological disaster and, in some cases, have caused irreversible environmental damage. The land is in extremely poor condition. There are numerous deep gullies; the bedrock is exposed where topsoil has been eroded away in some areas, and most of the natural vegetation cover has been lost. Every year 200,000 hectares of forest are lost. Most of the remaining forests are in the southern part of the country. West Wollega, Illu Ababor, Jimma, and Kaffa zones hold 53 per cent, while Arsi, Bale, and Sidama zones hold 38 per cent. The remaining forests are in North Shewa zone and a few other places. The northern highlands account for less than 4 per cent of the national forest cover.[6]

Before the establishment of Addis Ababa in 1886, dense afromontane forests covered the hills of Mount Entoto. The forests included such precious species as tid (*Juniperu sprocera*), weira (*Olea europaea*), kosso (*Hagenia abyssinica*), amiga (*Hypericum revolutum*), zigba (*Podocarpus falcatus*), grar (*Acacia abyssinica* and *Acacia negrii*), asta (*Erica arborea*), qega (*Rosa abyssinica*), and agam (*Carissa edulis*).[7] However, less than a decade after the new capital was established, the Entoto hills and their immediate surroundings were without trees, and any shrubs that could be used for fuel were also cleared. The situation was so severe that many in the early 1900s thought the new capital would have to be abandoned.

It was this critical situation that prompted Emperor Menelik II to issue decrees to resolve the wood shortage. His first decree was an attempt to prevent the burning of trees, but this order was largely ignored. His other decree, which prevented the cutting of trees without special permission, was too difficult to enforce. Following these failures, he took the suggestion of his foreign advisors and imported fast-growing new species of trees.[8] The country imported several species of euca-lyptus trees from Australia, of which *globulus* was the most suitable to plant in Ethiopia. Once imported, Menelik issued a proclamation in 1904 ordering the nationwide planting of this species.[9] The species – the *bahirzaf*, the tree that came from beyond the seas – grew very rapidly in the fertile, volcanic soils surrounding Addis Ababa, where precipitation is copious and fairly well spread. The extraor-dinarily quick growth and productivity of the *E. globulus* attracted the wealthier class to plant estates of trees for economic gain. Another advantage of the tree was that, when cut, it would grow back from its roots and could be harvested every two to three years. Imperial incentives also gave tax relief for land put into eucalyptus trees and the distribution of seedlings, resulting in the rapid spread of the tree. The eucalyptus tree saved Addis Ababa from being abandoned as the nation's capital. The tree was also responsible for the resurrection of other old wooded towns, such as Gondar and Adwa, and sparked similar developments in other towns.[10] Because the eucalyptus species lacked natural predators, the tree flourished wherever peo-ple planted it in the highland areas. Today, pure stands of eucalyptus trees enclose large and small urban agglomerations across the country. Menelik also started an afforestation scheme, but this project did not continue after his death.

During their occupation of the country in 1936, the Italians initiated the next forest resources and enforcement policy. The Italians resumed wide-scale deforest-ation by giving forests to private owners and allowing agriculture, urban encroach-ment, and other forms of forest destruction. Individuals and firms cut down timber to be processed at sawmills for export, construction, and furniture. Trees were felled all year round and the most valuable ones, such as cedar (*Juniperus pro-cera*), podo (*Podocarpus* spp.), tikur inchet (*Pygeum africannum*), and olive (*Olea* spp.), were chosen for extraction, leaving many forest areas in a poor state.[11] The Italians re-afforested the Entoto Mountains and Qabana River basin surrounding Addis Ababa with attempts at regulation.[12] However, efforts to improve or protect the forest resources of the country were hardly visible because of the ongoing countrywide war against the occupiers.

The post-independence 1943 Constitution of Ethiopia declared that the natural forest resources of the country were a 'sacred trust for the benefit of present and succeeding generations.'[13] All forest resources not owned by individuals were declared state-owned. The Imperial government stated its dedication to conservation and its commitment to preventing forest exploitation. The govern-ment introduced forestry education at the Ambo College of Agriculture that was established in the 1950s.[14] It added the Wondo Genet College of Forestry in the 1970s. The government also issued a series of proclamations in the 1960s: State Forest Proclamation, No. 225/1965; Private Forest Conservation Proclamation No. 226/1965, and Protective Forest Proclamation No. 227/1965. The objectives of these proclamations included protecting, conserving, and developing forest resources as well as protecting wildlife, regulating cutting, controlling forest fire, issuing leases and permits, ensuring the continuous supply of forest products, and reforesting to prevent soil erosion.[15] However, these laws were rarely enforced. Corruption was a problem, and in many cases, the very officials charged with forest protection cut down trees themselves for private gain and allowed their family members and friends to do likewise. Infrequently, the government would issue decrees against cutting trees on state lands and harshly punish those – mostly the poor – who were caught. The problem was especially acute in areas where provincial and local government officials were given free reigns to exploit the forest for their benefits. These officials would regularly allow illegal logging in their administrative areas and take a cut of the profits. Large areas of forest, especially those near watersheds, had no proprietor. In these areas, anyone could come and remove wood at will. Privately owned forests – estimated at 3 million hectares before 1974 – also suffered exploitation.[16] Some landowners considered forests to be impediments to crop cultivation and consequently had the land cleared. Others leased the land to commercial logging companies that stripped it of trees and left it open to erosion. Some species, such as the *zigba* (*Podocarpus garcilior*) and *tid* (*Juniperus procera*) conifers, which once covered much of the highlands, were removed to the point that they all but disappeared from the countryside. Only a few of these species survive in the remote areas of the central and southeast highlands.

The 1960s and early 1970s were a period of rapid and extensive deforestation. During this time, the demand for building materials, fuelwood, and charcoal rapidly increased, driven in part by urbanisation. For instance, the production of lumber doubled between 1955 and 1960, from 20,000 cubic metres to 40,000 cubic metres.[17] By 1960, annual production was estimated at 200,000 cubic metres for firewood (mostly *Eucalyptus globulus*), 20,000 cubic metres for posts and poles (mainly *Eucalyptus globulus*), 50,000 cubic metres for saw logs (mainly pinewood), and 60,000 cubic metres of wood (mainly acacia) for charcoal.[18] This increased demand for wood prompted a large number of wealthy people, such as government officials, to enter the logging business. Many of these people obtained pieces of state-owned forest, either by purchase at a nominal price or as pensions, gifts for service, or through bribery. Some, such as powerful aristocrats, simply enclosed portions of public land and claimed it as their own. Regions still covered by forest, such as Arsi, Bale, Sidama, Ilu Ababor, and Kaffa zones, became prime targets for these aristocrats. The government privatised tens of thousands of hectares of forestlands in these zones. At the end of the imperial era in 1974, absentee landlords owned three-quarters of the nation's forests.[19] Once privately owned, extensive logging followed.

During the time of Emperor Haile Selassie, there were some attempts to create an institutional framework for the purpose of environmental conservation. However, these measures were not implemented. The imperial government failed to initiate any programme to deal with the ongoing ecological degradation and deforestation until the early 1970s. It was only after the 1973–74 catastrophic drought and famine that the government realised the urgency of the problem, especially in the northern parts of the country. With the help of the World Food Program and other international organisations, the government launched several land rehabilitation programmes. It initiated extensive tree planting programmes were launched in North Shewa, South Wollo, and East Hararghe zones. It also implemented irrigation schemes, dam construction, and digging of wells and building water tanks in these other zones to combat drought. However, the overall effects of these efforts in rehabilitating the massive environmental deterioration that the country was faced with were minuscule.

The post-1974 revolution military regime also initiated a series of programmes to address the environmental problems of the country. In 1975, the Derg regime nationalised all land and forests fell under the control of the state. The government designated forests over 80 hectares as state forests, and their protection fell under the jurisdiction of the Forest and Wildlife Conservation and Development Authority. At the grass-roots level, the newly created Peasant Associations (PAs) were charged with conserving and managing community forests that were smaller than 80 hectares.[20] In 1980, the government issued Forest and Wildlife Conservation and Development Proclamation No. 192, which repealed previous forest proclamations and conservation orders. It provided for national parks, wildlife reserves, and sanctuaries to be registered, demarcated, and administered. The regime abolished private ownership of forests and spurred citizens to get involved in planting trees and conserving the country's forest and wildlife resources.[21] It also provided the Forestry and Wildlife Development Authority to carry out a countrywide survey of existing forests to initiate forest management programmes. The Survey led to the designation of 58 high forest areas in the country in the 1980s – covering a total area of about 3.6 million hectares – as National Forest Priority Areas (NFPAs). The purpose behind establishing these areas was to combat degradation and loss of the remaining forests.[22] However, the actual guidelines and rules regarding forest use were vaguely defined. Furthermore, only in a few cases was demarcation substantiated by forest management plans. Plans that were set up were often only rudimentarily followed.[23]

The government also established institutions and initiated several projects aimed at halting environmental degradation and restoring the land. It created new departmental units and subunits within the Ministry of Agriculture to combat deforestation and soil erosion. The State Forests Conservation Development Department (SFCDD) protected forests, closed degraded areas, and planted fuelwood around urban centres. The Community Forestry and Soil Conservation Development Department (CFSCDD) planned and operated afforestation and soil conservation projects in rural areas. These projects comprised three groups: farm forestry, community forestry, and soil conservation. The farm forestry projects encouraged farmers to plant trees – such species as *Cupressus lusitanica, Acacia decurens, Eucalyptus globulus*, and *camandulensis* – on their land for fuelwood and building materials. Through the PAs, the community forestry schemes provided technical and logistical support to communities, such as distributing seedlings to reforest and rehabilitate denuded areas on slopes greater than 35 per cent. The soil conservation projects encouraged farmers to take measures to conserve the soil by terracing, damming gullies, and building irrigation ditches. In the 1980s, the government also promoted large-scale peri-urban tree plantations around the large urban centres – including Addis Ababa, Bahir Dar, Dessie, Kombolcha, Gonder, Adama, and Debre Berhan – to meet the increasing demands for fuel-wood and timber for construction.[24] The government organised many of these programmes around food-for-work campaigns. The World Food Program (WFP) provided the food, while nongovernment organisations and bilateral government development agencies provided technical support and other services. Projects included 'construction of soil and stone bunds, check-dams, diversion canals, terracing, stream diversion, road construction, spring development, dam and pond construction, and seedling production and re-afforestation.'[25] These programmes did see results. By 1985, the PAs had planted some 500 million tree seedlings, built 700,000 kilometres (km) of terraces, put aside 80,000 hectares of steeply sloping land for natural re-growth, and covered 181,000 hectares of land with forests.[26]

While these results sound impressive, the scope, effectiveness, and sustainability of these environmental rehabilitation programmes were limited. This is due, in large part, to the inherent problems in the development strategy pursued by the military government. These programmes were guided solely by a rigid political ideology, and the populace often rejected such policies. While peasants had the right to cultivate parcels of land, lack of clear ownership had acted as a disincentive to farmers in terms of improving the land's productivity and participating in environmental rehabilitation programmes. Insecurity in land tenure rights discouraged peasant farmers from planting trees. They were less willing to plant trees and invest time caring for them, as they were not guaranteed to reap the benefits. Peasants working for WFP-supported afforestation schemes expected no other benefits for themselves beside the food that the programme compensated for their labour.[27] As a result, they showed little interest in caring for the seedlings they had planted. After all, food-for-work projects provided payment for planting trees but not for caring for the planted trees. Improper planting and poor maintenance resulted in a very low survival rate among the planted seedlings. Of the 500 million planted seedlings, only about 15 per cent managed to survive. It was especially true of trees planted on public land because the trees appear not to belong to anyone, and no one expects to benefit from the work. The government pursued re-afforestation without sufficient consideration of the ecological requirements of the tree species that were planted. For instance, eucalyptus seedlings were planted in areas with severe moisture deficiency and very shallow water table levels.[28] Overall, the government took a bureaucratic and dictatorial approach to environmental conservation. It valued peasants only for their labour. Local farmers were rarely consulted or included in the planning process. Government officials often discounted the experience of the locals and looked down on the peasants as being uneducated and lacking understanding of their environment. Farmers often have an intrinsic understanding of the ecology of the areas they cultivate.

Their experience would have been particularly useful for any environmental rehabilitation or conservation programme. Unfortunately, the regime's top-down, bureaucratic approach precluded any use of such local knowledge. As a result, the participation of the local farmers was undermined, and many of the projects failed to promote environmental stability.[29]

The protection of forests also never materialised because the management plans and designs failed to incorporate the participation and livelihood interests of people living within and adjacent to forests. The management system was top-down, unilateral, exclusionary, punitive, uniform, and rigid. In other words, the obdurate political ideology of the regime precluded the realisation of a management system that is decentralised, inclusive, participatory, enabling, sustainable, multipurpose, site specific, and dependent on local knowledge and institutions.[30]

In the 1980s, many fuelwood projects were implemented with support from multiple international donors. These projects were to alleviate fuelwood scarcity while preserving the country's forests. In 1989, two such projects were evaluated, both financed by the UN office for the Sudano-Sahel Program for environmental rehabilitation. In both cases, bureaucrats and forestry experts planned the projects centrally without regard for the peasants' interests. Experts had first estimated the fuel needs of two towns, Adama and Debre Berhan. Agricultural and forestry experts then identified hillsides near the towns that were only marginally agriculturally productive and demarcated this land for reforestation. The local peasant associations were instructed to mobilise peasant labour for the project. Many other towns had similar projects, including Addis Ababa. New forests on hillsides around the country sprang up.

However, as Siegfried Pausewang's study clearly indicated, the municipalities had to enforce harsh measures to protect these forests from the local population.[31] They placed restrictions on firewood collection, limiting supplies to local markets. Women who made a living gathering fuelwood were increasingly threatened by the forest guards, who would demand bribes, blackmail them, or even rape them. Theoretically, the fuelwood projects held some benefits for peasants. Peasants would either be paid outright or through food for work programmes for their support. After they plant the trees, peasants would further benefit from the sale of timber and fuelwood. However, the PAs organised peasant participation in a way that discounted their interests. Peasants were told to hand over land 'voluntarily.' Administrative leaders ignored peasants' complaints; when peasants protested the loss of grazing land, they were only told to take their cattle elsewhere. Peasant labour was used to plant forests, and the government kept these forests under state control until the peasants were ready to take over their management. For those employed in the nurseries or administrative jobs, there were some profits. Other peasants benefitted through food-for-work programmes or other payments. Additionally, peasants were told that once the forests were ready for harvesting, they would profit from the sale of the wood. Despite such promise, peasants' complaints increased. The loss of grazing and agricultural land was especially hard for peasants to bear. The peasant associations promised compensation. However, there was simply no land to spare for redistribution, and so, such compensation never materialised. In retaliation, peasants stole wood from the forests, grazed their cattle there, and sometimes cut down trees.[32]

After removing the Derg regime from power in 1991, the Ethiopian People's Revolutionary Democratic Front (EPRDF) government continued the commitment to public ownership of land. The state became the owner of all forests. The government issued proclamations that set guidelines for the conservation, development, and use of the forests: the Proclamation to Provide for the Conservation, Development, and Utilization of Forests No. 9/1994. This regulatory framework repealed all previous forestry legislation and is still in force today. The proclamation was to preserve the ecological stability of forests, including the protection of rare and endangered animal or plant species, to prevent soil erosion, and to utilise forest resources for development. It revised forest policy and marked a major change in forest ownership, tree tenure rights, and the marketing of forest products. Specifically, it promotes the involvement of the private sector and recognises that communities living in or near the forest should benefit from its development.[33] The proclamation also responded to regionalisation by giving more authority to the regional states. Following decentralisation and ensuing delegation of power, regional states were mandated to manage forest priority areas in their administrative jurisdictions. In a major forest policy shift, the proclamation encouraged private tree planting for personal use or profit or both. Peasant farmers were encouraged to grow trees on their farms for their consumption and market if they wished to do so.[34]

The 1994 proclamation separated forests into five administrative categories. *State Forests* are those 'forest areas that the federal body has designated, demarcated, and registered to develop forest resources, protect genetic resources, and conserve the ecosystem.' *State Protected Forests* cover forestland designated, demarcated, and registered by a federal body for the purpose of conserving the ecosystem and genetic resources by protecting it from human or domestic animal encroachment. *Regional Forests* are designated, demarcated, and registered by the regional governments to conserve the ecosystem and genetic resources. *Regional Protected Forests* are designated, demarcated, and registered by the regional government to protect the ecosystem and genetic resources by preventing the encroachment of humans or domestic animals. Any legal individual, an entity such as a PA or a group of individuals or local communities, develops *Private Forests*. Regional governments register and administer private forests within their regional boundary.[35]

The 1994 legislation prohibited the cutting of trees, grazing of domestic animals, hunting, beekeeping, and harvesting of other forest products in protected forest priority areas without written authorisation from federal or regional authorities. The rules strictly prohibited the removal of such indigenous tree species as *Juniperus excelsa*, *Cordia*, *Podocarpus falcatus*, and *Hagenia abyssinica* from regional and state forests. Unauthorised activities in these forest areas carried a fine of up to Birr 500 or a maximum of two years in prison.[36] Federal and regional authorities carried out their mandatory obligations with little or no participation by local communities. Despite legislative pronouncements regarding benefits sharing with local communities, people living in and around existing forests were kept out of decision-making about forest protection and accessing forest resources. The legislation failed to empower local communities to manage forests now under the national and regional jurisdictions. Consequently, the loss of forests continues because of the lack of local control and ownership as well as the government's lack of capacity to provide security. An estimated 113,000 hectares of forests are destroyed annually.[37] According to the Ethiopian Environmental Protection Authority, three-fifths of the total surface of the 58 NFPAs is degraded, and nearly a third of the same total surface area has suffered severe degradation.[38] Twenty-four of the NFPAs have no high forests.[39]

In 2007, however, the government issued yet another forest development, conservation, and utilisation policy that relegated forest management to regional governments and local communities. Regional governments were empowered to manage nearly all of the NFPAs that are located within their jurisdiction. Regional governments were obligated to promote community-based forest management as a means for forest conservation and protection with government and non-government expatriate organisations providing financial, technical, and organisational assistance.[40] In this regard, the new policy stipulated a series of regulations. First, forest development, conservation, and utilisation plans shall be formulated to allow local communities to participate in the development and conservation and share benefits from the development of state forests. Second, the local community may utilise forest products from the state forest to the extent necessary for satisfying their ordinary domestic needs in accordance with directives to be issued by the appropriate regional body and in conformity with the management plan developed for the forest. Third, the harvesting of forest products, grass, and fruit, as well as the keeping of beehives in state forests, may be allowed based on the objective realities of the locality. Finally, the local community may be allowed to keep beehives and produce spices, forest coffee, and forage in a protected forest by providing training and technical support.[41] This new policy seems to guide forest management in the right direction, but a lot remains to be done in terms of involving local communities to ensure the sustainability of what is left of the country's forest resources.

The fate of the remaining forests and woodlands

Deforestation is occurring in Ethiopia, but no one knows the exact rate or extent of it. The Ministry of Agriculture estimates the annual deforestation rate at 163,000 hectares. *Earthtrends* estimates a loss of 400,000 hectares of forest between 1990 and 2000. The Ethiopian Forestry Action Plan estimates between 150,000 and 200,000 hectares per year. More recent estimates show that forest cover declined from 3.6 million hectares in 2000 to 3.3 million hectares in 2005.

The high woodland area declined from 10 million hectares to 9.6 million during the same period. Regardless of which estimate is factual, the loss is huge.[42]

The remaining forest resources of Ethiopia cover 27.5 million hectares of the land, of which 2.3 million hectares is considered high forests, 5 million hectares woodland, 20 million hectares bushland, and the remaining 0.2 million hectares plantation.[43] Most of the existing high forests are in the southwestern and south-eastern parts of the country. The occurrence of high and reliable rainfall in these regions has produced dense forest cover. In the 1980s, high forests and some dense woodlands were identified and classified as National Forest Priority Areas (NFPAs) for management and protection purposes. The 58 NFPAs cover about 3.7 per cent of the land surface of the country. Their sizes range between 10,000 to 300,000 hectares. Over 300 plant species are found in these forests.[44] The NFPAs have boundary demarcation, but few have legal protection. The administration and management of these forests have recently come under the jurisdiction of the regional state governments. Local communities living inside and outside NFPAs are legally obligated to assist local government authorities in protecting the forests from illegal activities, but the authorities rarely seek their inputs in managing the forests.[45] The involvement of the local communities is vital in the management of these forests.

Forests in the Oromiya regional state

More than three-fifths of the remaining high forests in the country are found in the Oromiya regional state. A little over half a century ago, Oromiya had an esti-mated 14.7 million hectares of high forests, covering some 40 per cent of the state. Today, high forests cover only 7 per cent of the land. Woodlands and bushlands in Oromiya include an estimated 12 million hectares (32.7 per cent) of the total area of the state. Woodlands primarily include *Juniperus*, broadleaf deciduous *combretum terminalia, acacia*, and *acacia commiphora*. The estimate for annual forest clearing in Oromiya is 1.2 per cent.[46]

Since the late 1990s, Oromiya has been involved in extensive re-afforestation and land rehabilitation programs. The regional state has promoted industrial plan-tation within and around regional forest priority areas in Arsi, Borana, Illu Ababor, Jimma, East and West Wollega, East Shewa, and West Shewa zones. Peri-urban plantations are also promoted to meet the growing demand for construction and fuel wood in urban areas. *Eucalyptus* spp. and *Cupressus lusitanica* are the favoured species for the peri-urban and industrial plantations. The region promotes agroforestry or farm forestry practices. Households grow *Eucalyptus* spp., *Cordia africana, Croton macrostachus, Rhamnus prinoides* (Gesho), and various species of fruit trees around farm dwellings. Indigenous tree species, such as *Faidherbia albida, Cordia Africana, Croton macrostachus, Albizia lebeek, Mellitia*, and *ferrugenia*, are retained or planted on cultivated fields. These trees provide fod-der, wood for construction and fuel, and nutrients to the soil. Farmers are also encouraged to devote a piece of their farmland to grow trees. Farmers seem to be willing to do so because they now have life-long, inheritable rights to the land they cultivate. By and large, *Eucalyptus globulus, Eucalyptus saligina*, and *Eucalyptus camaldulens* are grown on such farm woodlots. The regional government pro-vides technical and material assistance to farmers and communities to plant trees on farm boundaries and roadsides and as windbreaks. They are also promoting afforestation of communal land and eroded hillsides. Overall, about 2.2 billion tree seedlings were produced and planted in the state on 205,000 hectares of land during two consecutive five-year development plans that ended in 2004.[47]

On a larger scale, the Oromiya regional state is responsible for managing the NFPAs located within its jurisdiction. The state is home to 37 of the country's 58 National Forest Priority Areas, now referred to as Regional Forest Priority Areas (RFPAs). While the regional government manages most of these forests, some of them are designated as community forests. The Oromiya Forest Proclamation No. 72/2003 supports 'the participation of local communities living within and adjacent to state forest priority areas in the conservation, development and proper utilization of state forests.'[48] However, the law does not clearly define the nature and extent of such participation.

Despite the state's efforts to manage and protect the remaining forests, the use of forest resources and lack of enforcement of prohibited activities in forest priority areas continue to be tolerated. For example, take the case of the Munessa-Shashamane State Forest that straddles between East Shewa and Arsi zones. This tract of forest, located on the eastern shore of Lake Langano, is officially recognised as a nature reserve. It also covers the steep-sided eastern escarpment of the Rift Valley overlooking the lake. This 98,000-hectare forest is predominantly *Podocarpus gracilior*, *Junipers procera*, *Pygeum africanum*, *Ficus sur*, *Cordia abysinica*, and *Ficus sycamorous*. In the past few decades, farmers have moved down the eastern Rift Valley escarpment toward the eastern shore of the lake. Much of this forest has significantly been reduced by the expansion of agricultural and grazing activities. Ample precipitation (1,200 millimetre (mm)/year) and clay loam soil that dominate the areas allow farmers to produce wheat, barley, maize, sorghum, and varieties of legumes.[49]

The Jimma Zone in the western Oromiya regional state is also relatively endowed with forests. Its 85,000-hectare Tiro-Botor-Becho forests are fairly well managed.[50] The forests have shallow soils and deep slopes, making logging and cultivation of crops difficult. The state has proposed this forest area as a biodiversity reserve. It is a mixed coniferous forest with broad-leaf species, including *Aningeria adolfi-friderici*, *Apodytes dimidiate*, *Olea capensis*, *Syzygium guineense*, and, to a lesser extent, *Podocarpus falcatus*. *Juniperus procera* mixed with *Hagenia abyssinica* occurs at elevations above 2,500 metres and *Erica arboreal* above 3,000 metres.[51] The 174,000-hectare Gera forests in the same zone are likely the largest continuous forests in the region, have a significant amount of mature timber, and are heavily underlain with coffee. Increased development in recent years is threatening these forests, however. For instance, the coffee road that was built through the forest has encouraged both increased coffee production and cultivation of food crops. Overlooking the Gojeb gorge is a narrow 74,000-hectare forest area that is too steep for cultivation. These forests are crucial watershed areas.[52]

Other major high forest areas in western Oromiya are in the Illu Ababor zone, including: Sigmo Geba (280,000 hectares), Abobo-Gog (218,000 ha), Gebre Dima (165,000 ha), Godere (160,000 ha), Sele-Anderacha (225,000 ha),

Sibo-Tale Kobo (100,000 ha), Yayu (150,000 ha), Yeki (122,000 ha), Wangus (330,000 ha), and Mesengo (292,000 ha).[53] In 2010, the Yayu forest was declared one of the world's vital Biosphere Reserves under UNESCO's Man and the Biosphere program. This area

> comprises undisturbed natural forests and semi-forest systems managed for the production of coffee, spices, honey and wood while providing important ecosystem services such as watershed protection in the Nile Basin. The Yayu forest has the greatest abundance of wild *Coffea arabica* anywhere. Sustainable development activities in the site focus on coffee production, including the planting of fruit trees both for crop and to provide shade for the coffee.[54]

Over half a million hectares of land (Saylem Wangus, Sibo Tole Kobo, and Debre Dima forests) are also shared between the Oromiya and Southern Nations regional states. These forests are considered high forests with continuous closed canopy cover but have been seriously changed by humans. Other forests have been cleared for coffee and tea plantations.[55] Unfortunately, over 38 per cent (782,000 hectares) of these forests are considered heavily disturbed by anthropogenic activities.[56]

In central Oromiya, about 40 kilometres west of Addis Ababa is the 9,800-hectare Menagesha State Forest. It is on the southwestern slopes of Mount Wechecha (3,385 metres above sea level). The area surrounding the forest is intensively farmed and grazed. *Juniperus procera* intermixed with *Erica arborea*, *Rosa abyssinica*, and the endemic *Jasminum stans* are found at higher elevations between 2,500 m and 2,900 metres above sea level, while *Olea europaea cuspidata*, *Allophylus abyssinicus*, *Maytenus* spp., and *Euphorbia ampliphylla* form the understory. *Podocarpus gracilior* spp. are also scattered throughout the lower elevations. Giant herbs like *Lobelia gibberoa* and *Solanecio gigas* are ubiquitous on the sides of the valley, while the *Scadoxus multiflorus* covers the forest floor.[57] Various species of pine trees and eucalyptus have been planted as a buffer in the lower areas of the forest via reforestation efforts since the 1980s.

The Menagesha State forest is a habitat for three Ethiopian endemic bird species: Yellow-fronted Parrot, Abyssinian Woodpecker, and Abyssinian Catbird. Other bird species recorded include Abyssinian Black-headed Forest Oriole, Abyssinian Ground Thrush, African Hill Babbler, Black-winged Lovebird, White-cheeked Turaco, Slender-billed Chestnut-winged Starling, Sharpe's Starling, and Bended Barbet. Colobus Monkey, Serval, Grey Duiker Warthog, Porcupine, and Civet Cat are also known to inhabit the forest.[58] One major concern for the conservation of this forest is insufficient protection; illegal felling of trees for construction and fuel continues unabated, despite re-afforestation efforts by the Menagesha Forestry Training Center.

Addis Ababa's northern limit is Mount Entoto, which rises to over 3,200 metres. The hills of Mount Entoto are a National Park. Established in 1995, it covers about 13,000 hectares of forestland. Dense afromontane forests that included such precious species as *Juniperus procera, Olea europaea, Hagenia abyssinica, Hypericum revolutum, Podocarpus falcatus, Acacia abyssinica* and *Acacia negrii, Erica arborea, Rosa abyssinica,* and *Carissa edulis* once covered the hills. These forests had full of life. Few of these indigenous species remain today, as fast-growing eucalyptus species are prevalent, and little life dwells beneath the canopy of these trees. The quickly growing eucalyptus species have 'suppressed the growth of nearly all indigenous' species, and only *Juniperus procera* has survived in proximity to the eucalyptus stands.[59] Eucalyptus trees now cover most of the hills of Entoto. Local environmental groups are striving to replace eucalyptus with indigenous plants. The groups argue that eucalyptus is hazardous for the environment because the species neither prevents soil erosion nor restores soil fertility. They believe that restoring the hills to more native species such as *Juniperus excelsa, Acacia abyssinica, Olea europaea cuspidata,* and *Hagenia abyssinica* will reduce soil erosion and flooding and can help bring back wildlife to these picturesque hills. The Ethiopia Heritage Trust (EHT) has been the chief organisation in the forefront of re-afforestation efforts since the early 1990s. There is some evidence that wildlife is returning to the reforested sections of the hills, including mammals like Civet Cat, Spotted Hyena, Grey Duiker, White-tailed Mongoose, and monkeys.[60] The Ethiopian Wildlife and Natural History Society (EWNHS) has recorded 115 bird species in the Park, including endemic to rare birds like Rouget's Rail, Wattled Ibis, Abyssinian Slaty-flycatcher, Abyssinian Catbird, Thick-billed Raven, White-backed Tit, Spot-breasted Plover, Yellow-fronted Parrot, Black-winged Lovebird, and Abyssinian Woodpecker.[61] The expansion of the city up the hills of the mountain range and unsustainable harvesting of the forest for firewood remain major threats to the integrity of the Park and its ecosystem.

The Harenna forests (100,000 hectares) in eastern Oromiya are the second largest continuous montane moist forests remaining in the country. These forests are located high on the Bale Mountains, which rise to the afro-alpine Sanetti Plateau at 4,300 metres above sea level. To the south of this plateau are the Harenna forests, which range from 3,500 metres to 1,500 metres at the park's southern border. The greater Harenna forest comprises five different Regional Forest Priority Areas as well as the Bale Mountains National Park (BMNP). Mana Angetu is the largest regional forest priority area in Harenna, with forests covering an area of 180,584 hectares.[62] Annual precipitation exceeds 1,000 mm per year. Rainfall is bimodal. Between March and April, there is a short rainy season followed by a longer one from August to October.

The Harenna forests are considered to be one of Ethiopia's biological hotspots. It has a wide range of plant species, some of which are not found anywhere else in Ethiopia. Two such species are *Filicium decipiens* and *Alangium chinense.* The Harenna forests are also important for its *Coffea arabica* population. Coffee in the Harenna forests is a dominant understory shrub and shows a high degree of physiological variability.[63] The forest vegetation includes the upper canopy, sub-canopy, shrub, and ground layers. The upper canopy is made up of emergent trees of *Pouteria adolfi-frederici,* with *Afrocarpus falcatus* and other species such as *Olea capensis* sub-sp. *welweitschii* and sub-sp. *hochestetteri, Prunus africana, Albizia schimperiana, Milletia ferruginea, Celtis africana, Polyscias fulva, Schefflera volkensii, Trilepisium madagascariense, Schefflera abyssinica, Bersama abyssinica,* and *Mimusops kummel.* The species of the sub-canopy include *Croton macrostachyus, Cordia africana, Dracena steudenri, Syzygium giuneense* sub-sp. *afromontanum, Sapium ellipticum, Ilex mitis, Erythrina brucei,* and *Rothmannia urcelliformis.* The shrub layer consists of *Coffee arabica, Galiniera saxifraga, Teclea nobilis, Ocotea kenyensis, Clausena anisata, Measa lnceolata* and *Maytenus* spp. Woody climbers include *Urea rahypselodendron, Landolphia owarensis, Embelia schimperi,* and *Jasminum* spp. The herbaceous plants of the ground layer are *Acanthus, Justicia, Piperoma, Galinsoga, Impatiens, Urtica,* and several species of grass. This biome also supports epiphytes *Canarina, Orchids, Scadoxus,* and ferns such as *Platycerium* and *Drynaria,* as well as various mosses in the wettest portions of the forests.[64]

The Adaba-Dodola is another 73,000-hectare forest priority area located in the northern slopes of the Bale Mountains that serves as a buffer zone for the Bale Mountains National Park. Conifers such as *Juniperus excels* and *Podocarpus falcatus* sp. dominate areas with elevations under 3,000 m above sea level, while *Hagenia abyssinica, Hypericum lanceolatum,* and *Erica arborea* overshadow other species at elevations exceeding 3,000 m above sea level. Other indigenous tree species found in this forest include *Olea europea ssb subsidata, Alophylus abyssinicus, Nuxia congesta, Ekebergia capensis, Scheffleria abyssinica, Buddleya polystachia, Mytenus* sp., and *Pittosporum viridiflorum. Eucalyptus* sp. and *Cupressuslusitanica* are grown in the peripheral areas where logging has removed old native species.[65]

The forests of the Bale Mountains have been spared from major destruction because of their remoteness. Many of these forests are far from cities and, thus, are not easily accessible. Transporting timber and forest products to manufacturing sites is a long and expensive process. But as human settlements increase, farms and grazing lands also expand, threatening the forests. People harvest timber and non-timber forest resources unsustainably.[66] A total area of over 202,000 hectares of forest is considered heavily disturbed in the Mena Angetu, Kubayo, Harenna-Kokoss, Goro Bele, and Logo forest priority areas.[67] Conflicts have arisen between local communities and state authorities over these forests. Local communities want to pursue their traditional way of life by accessing forest resources, while the regional state government wants to protect these remaining forests from being destroyed. State authorities have forcibly removed and relocated many communities from the forests. Unsurprisingly, local communities are resentful and keep returning to their homes inside the forests.[68]

Uncontrolled fires also represent a serious threat to the Bale Mountains forests. Forest fires have become more recurrent in the region in recent years. In the early 1990s, large sections of the Mena Angetu forest priority area were destroyed by wildfires.[69] A series of forest fires also broke out in 2000 in several parts of the country, including Bale, Borana, Jimma, Illu Ababor, East Wollega, East Hararghe, and Arsi zones. Hundreds of thousands of hectares of forests burned down, including many of the NFPAs. In Bale and Borana alone, between 100,000 and 150,000 hectares of forests went up in flame.[70] For days, local communities watched their forests burn because neither the regional state nor the federal government possessed adequate human-power, technical know-how or logistical capability to fight forest fires. Massive destruction of the forests was prevented only after a fire brigade, which came all the way from South Africa, provided assistance and put out the fire in a relatively short time. The causes of forest fires in the region are both human-made and natural. Lightening causes some of the fires, but human agents are the principal culprits. People start forest fires, but most are accidental. Accidents usually happen while clearing forest land to make room for farming, burning the grass to induce re-growth, controlling weed infestation, producing charcoal, or smoking out bees from beehives to collect honey.[71]

The destruction of the Bale Mountains forests will have many repercussions. These forests are the most ecologically diverse in the country and are home to numerous plant, animal, and avifauna species, many of which are endemic. *Warburgia ugandensis*, *Filicium decipiens*, *Prunus africana*, *Hagenia abyssinica*, and *Afrocarpus falcatus* are currently endemic and threatened in the region. Large mammals that are under threat include: Mountain nyala (*Tragelaphus buxtoni*), Menelik's bushbuck (*Tragelaphus scriptus meneliki*), Bale monkey (*Cercopithecus djamdjamensis*), Bohor reedbuck (*Redunca reduncabohor*), Serval (*Felisserval*), Guereza colobus monkey (*Colobus guereza guereza*), African wild dog (*Lycaon pictus*), Golden jackal (*Canis aureus*), Lion (*Panthera leo*), and the Giant forest hog (*Hylochoerus meinertzhageni*). Endemic and threatened bird species include the White-backed tit (*Parus leuconotus*), Abyssinian catbird (*Parophasma galinieri*), Abyssinian woodpecker (*Dendropicos abyssinicus*), Black-winged lovebird (*Agapornis taranta*), Spot-breasted Plover (*Vanellus melanocephallus*), Blue-winged goose (*Cyanochen cyanoptera*), Red-billed chough (*Pyrrhocorax pyrrhocorax*), Golden eagle (*Aquila chrysaetos*), Bale parisome (*Parisoma griseaventris*), and Ruddy shellduck (*Tadorna ferruginea*). There are also endemic rodents that will be lost, including the Giant molerat (*Tachyoryctes macrocephalus*), Blick's grass rat (*Arvicanthis blicki*), and Harsh-furred rat (*Lophuromys melanonyx*).[72] Thus, the deforestation of these forests can set off 'change in the animal community by eliminating habitats and food resources, causing the extinction of host plants, decreasing habitat heterogeneity, and creating forest fragments that restrict animal movement.'[73]

Bordered by the Genale River to the east and Awata River to the west, the Anferara-Wadera forests are located in Oromiya's northern Borana zone. The 106,568-hectare Regional Forest Priority Area represents most of the high altitude forests in Southern Ethiopia. The forests cover a highly rugged and broken topography that is unsuitable for agriculture and grazing. These forests are heterogeneous. *Podocarpus falcatus* dominates the northern part of the forest along with broadleaf species, such as *Croton macrostachyus*, *Hagenia abyssinica*, *Ilex mitis*, *Olea capensis*, *Schefflera abyssinica*, and *Syzygium guineese afromontanum*. *Aningeria adolfi-friderici* species cover much of the southern part, where precipitation is higher. This species is the 'tallest and most important tree of these forests. The canopy below comprises *Albizia gummifera* and other *Albizia* spp., *Celtis africana*, *Ekbergia capensis*, *Fagaropsis angolensis*, *Ocotea kenyensis*, *Olea capensis*, *Phoenix reclinata*, *Polyscias fulva* and *Prunus africana*.' As one gets nearer to Negele, the vegetation turns to dry montane forest that was once dominated by *Juniperus procera*, but few of these species live today. Scrubs and tree species, such as *Barbeya oleoides*, *Catha edulis*, *Olea europaea cuspidata*, *Pistacia aethiopica*, *Pittosporum* spp. and *Schrebera alata*, have taken over.[74]

Although the government designates the Anferara-Wadera forests as National Forest Priority Areas (NFPA), they are far from being conserved. According to the Ethiopian Wildlife and Natural History Society and BirdLife International, the Anferar-Wadera

> forests are currently one of the main sources for timber of *Aningeria adolfi-friderici* and *Podocarpus gracilior*, which are being harvested with sawmills in the forest. Although some attempts to introduce sustainable harvesting methods have been made on paper, in practice the areas harvested are usually clear-cut, leaving insufficient cover for the regeneration of tall forest trees, particularly *Aningeria adolfi-friderici*. The primary threat to the area is the rapidly increasing (8.3 percent per year) human population living within the forest. These people make a living from cutting trees for charcoal, timber or fuel. They also clear the forest for cropland and grazing. Honey production has also been destructive: large areas of forest have burned when people use fire in the hives during honey collection. Open-cast gold mining poses an additional threat to the ecosystem.[75]

Forests in the SNNP regional state

Nearly a fifth of the remaining high forests in the country are in the SNNP regional state.[76] These forests comprise broadleaf, mixed, riverine, and plantation forests. Most of the natural forests are in Kaffa, Bench Maji, Shaka, and Dawuro zones and Konta Special District. These districts have some of the highest rainfall in the country. Plantation forests are most predominant in the Gamo Gofa, Gedeo, Gurage, Hadiya, Kembata, Timbaro, Sidama, Silte, and Wolayita zones and Alaba Special District.[77] The Kaffa Zone is the most extensively forested in the region. Nearly a third of the zone has dense forests. The zone accounts for more than one-half of the country's remaining Afromontane evergreen forests ecosystems.[78] They are high primary forests with a significant degree of biodiversity. Well over 100 plant species have been identified in these forests alone. *Coffea arabica* originated in the Kaffa highland forests. These forests still contain naturally regenerating populations of wild plants.[79] These forests are major sources of livelihood for local communities. For instance, nine out of ten households in Bench Maji, Kaffa, and Shaka zones collect forest honey, coffee, spices, and other non-timber forest products (NTFPs) from neighbouring forests. These resources account for one-third to three-quarters of the household's annual income.[80]

Located in the central Kaffa zone, the Bonga forests are the largest patches of forests in the region. The Bonga forests are among the 58 NFPAs designated for the protection of biodiversity and the conservation of resources. They are now designated as Regional Forest Priority Areas. The forests cover, wholly or partially, five districts, including: Decha, Chena, Gimbo, Menjiwo, and Tello. The Bonga RFPA covers an area of about 162,000 hectares but only one-half, or 81,000 hectares, constituted a high forest. Woodland, plantation, grassland, and cultivated lands cover the remaining area.[81] Precipitation in the region is copious, the mean annual rainfall is between 1,700 mm and 2,000 mm. Three major rivers – the Dincha, Gojeb, and Woshi – and their numerous tributaries drain the region. The Bonga forest has three distinct vegetation types based on elevation. Upland Rain Forest Vegetation occurs between 1,500 m and 2,200 m elevation. The dominant tree species in this zone include *Olea welwitschii, Scheffleria abyssinica, Euphorbia obovalifolia, Croton macrostachyus, Albizia schimperiana, Prunus africana, Syzygium guineense*, and *Polyscias fulva*. This zone also contains smaller trees and shrubs, such as *Milletia ferrugina, Teclia nobillis, Dracaena steudneri, D. afromontana, Galiniera saxifrage*, and herbs include false Cardamom (*Afromomum corrorima*). Upland Humid Forest Vegetation occurs between 2,450 metres and 2,800 metrees altitudes and contains tree and shrub species, such as *Apodytes dimidiate, Brucea antidysenterica, Barsama abyssinica, Coffea arabica, Clausena anisata, Erythrina bruci, Galineiera saxifrage, Hagenia abyssinica, Ilex mitis, Milletia ferruginea, Myrsine melanophloeos* (*Rapaenia simensis*), *Oxyanthus specious, Phonix reclinata, Pittosporum viridiflorum, Psychotria orophila, Rothmannia uricelliformis, Solanecio mannii, Oncoba routledgei, Enset ventricosa*, and *Chionanthus mildbraedii*. Pure stands of bamboo thickets or bamboo mixed with *Hagenia abyssinica, Myrsine melanophloeos*, and *Hypericum revolutum* occur between 2,400 metres and 3,000 metres elevation.[82]

Wildlife in and around the Kaffa forests includes: *Vivera civeta, Silvica pragrimmia, Papio Anubis daguera, Syncerus caffer, Traglaphus scriptus, Patomochoerus porcus, Cercopithecus mitis, Colobus guereza, Genetta rubiginosa, Viverrdae sanguineus*, Ethiopian hare, Jakal, Hyaena, Felis pardus, Felis leo, Hystrix cristata, Rock python, Snakes (black, red and striped), *Phacochoerus africanus*, and *Caris simensis*. Bird species include *Bostrichia caruculata* (Watttled Ibis), *Cyanochen cyanoptera* (blue winged goose), *Poicephalus flavifronse* (Yellow- fronted parrot), *Parophasma galinieri* (Abyssinian cat bird), *Parus leuconutus* (White-backed black tit), *Onchoganthus albirotris* (White-billed starling), *Oriolus monanacha* (Black-headed forest oriole), and *Caruvus crassirostris* (thick-billed raven).[83]

Forest depletion is about 2.4 per cent, or 80,000 hectares, annually, in the SNNP regional state.[84] Anthropogenic factors are mostly responsible for the increasing loss of forests in the area. The most important of these are the expansion of agriculture, urban development, road construction, and resettlements. People from the northern highlands have migrated to the area in the past century, mainly due to land alienation and coffee development. Over the last four decades alone, successive governments have also encouraged settlement of famine-stricken people from the north into this area, where precipitation is higher and more reliable and the land is more productive. Historically, most people in the area relied on forest resources supplemented by the cultivation of enset and root crops. They collected wild coffee, honey, spices, grass for thatching, and animal feed from the forests. Many hunted in the forests as well. However, the settlers brought with them new agricultural practices, and the region's economy gradually shifted from harvesting forest resources to tillage farming and cultivation of grains, such as maize and tef. Plow cultivation expanded because of the ability to farm larger areas of land with less labour. With an increased settler community, coffee production was also encouraged and expanded by the state in order to raise revenue, making the region one of the primary coffee producers and an important resource area for the central government. More than a century of central government and settler-induced development activities in the region has resulted in many non-sustainable land uses. For instance, increased agricultural production demands necessitated forest clearance. The increased population density and pressure on the natural resource base interfered with existing fallowing systems and the use of resources, especially forest resources. State coffee and tea farms were constructed by removing vast amounts of forests. Inadequate and corrupt administration of timber extraction licenses leads to damaged forests and a lack of re-growth measures.[85]

The growth of urban settlements in the region also exacerbates deforestation. For example, urban areas, such as Gimbo, Bonga, Decha, Wushwush, Kobech, and Saja, surround the Bonga forests. The demand for wood for fuel and construction is growing with the increasing populations of these towns.[86] Hardy species, such as *Cordia Africana*, *Pouteria adolfi-friederici*, and *Prunus africana*, are specifically removed for construction material from the Bonga, Boginda, and Mankira forests. As a result, these species are fast dwindling in their numbers.[87] The construction of roads – Jimma-Bonga and Bonga-MizanTeferi, for example – has increased the rate of forest depletion. Improved accessibility has also induced forest fragmentation as human settlements are interspersed in forest areas. Forest fragmentation has, in turn, increased deforestation activities.[88] Overall, 65 per cent (153,000 hectares) of high forests in the Kaffa zone are considered heavily disturbed.[89]

Over the last decade or so, the regional state government has been attempting to protect the remaining forests in the region by promoting participatory forest management programmes in surrounding communities. At present, there are 60 local participatory forest management groups (PFMGs) in Kaffa managing over 130,000 hectares of natural forests.[90] PFMG members can harvest NTFPs, such as honey, spices, and wild coffee in the protected forest. In the buffer zones around the protected forests, people also cultivate coffee, cardamoms, long pepper, and fruit on their plots of land.[91] Re-afforestation is also a significant environmental activity where community groups have planted tens of millions of seedlings on the region's degraded areas as part of watershed rehabilitation programs.

In 2010, UNESCO's International Advisory Committee for Biosphere Reserves recommended that 760,000 hectares of the Kaffa forests be a Biosphere

Reserve under UNESCO's Man and Biosphere Program. The recommendations accord one-half of these forests full protection. while the remaining half is to support local sustainable livelihood endeavours.[92] The Berlin-based Nature and Biodiversity Conservation Union (NABU), which proposed the Kaffa forests for Biosphere Reserve status, is assisting communities around the forests by providing funding for reforestation and sustainable coffee management projects and distributing improved wood stoves to help reduce deforestation.[93] More individual and collective efforts are needed to conserve these forests. Their sustainability is crucial to the economy of the region because a significant proportion of the inhabitants' livelihoods depends on them.

Forests in the Gambela regional state

The Gambela regional state is home to about 13 per cent (nearly half a million hectares) of the remaining high forests in the country.[94] Ample precipitation and the Baro, Gilo, Itang, Alwero, and Akobo rivers make the region the wettest in the country. The eastern part of Gambela comprises the catchments of the Baro and Akobo Rivers and is endowed with moist evergreen lowland forests, including the Godere, Yeki, and the Akobo-Gog forests. These forests are at altitudes between 450 metres and 1,400 metres with the wet season lasting up to eight months and the annual rainfall ranging between 1,300 mm and 1,900 mm.[95] A vast area of the Gambela state has been designated as a protected conservation area, including the Gambela National Park and the controlled hunting areas of Jikawo and Tedo. The Gambela National Park and the forests along the Baro River are home to near threatened species as Shoebill, Black-winged Pratincole, Basar Reed Warbler, and large numbers of water birds. The region has as many as 300 bird species.[96]

Although the region is less densely populated and is the most inaccessible in the country, it is not immune from losing some of its high forests. The rate of forest loss is about 1.3 per cent annually.[97] People in the state – whether rural or urban dwellers – depend on wood for construction and cooking fuel. Roads are being constructed, and large-scale irrigation agriculture is planned to develop the state. These and other future developments will undoubtedly produce harmful impacts on the forests and wildlife of the state. Occasional forest fires also cannot be ruled out.

Forests in Tigray and Amhara regional states

Not much natural vegetation remains in the northern highlands of Ethiopia. Forests cover less than 200,000 hectares in the Amhara and Tigray regional states combined. Indigenous species like *Olea Africana* and *Juniperus procera* are rare. One finds patches of woodland around churches and remote mountains. Whatever natural forests did exist have mostly been removed to make room for subsistence farming. However, a few remnants of indigenous forests are found in scattered remote locations in the northern and central highlands of the country. The Desse'a state forest is one of them. It is located on the edge of the escarpment in Eastern Tigray and covers about 120,000 hectares of land. Parts of the forest suffered during the civil war of the 1970s and 1980s as war-ravaged local communities were forced to cut down the forest for firewood and sell it in nearby towns, such as Mekelle and Quiha. After the civil war ended in the early 1990s, management of the forest was transferred to the regional state's Bureau of Agriculture. The Bureau put in place regulations governing the public use of the forest, including forbidding the cutting of live trees and permitting the collection of dry wood for home consumption but not for sale in the market. However, these rules were hard to enforce, as the communities surrounding the forest objected to the prohibition on collecting dead and dry timber, the total ban on cutting live trees to make farm implements that are critical to agricultural activities.[98] The sustainability of this forest would be enhanced if managed by the local communities for the simple reason that 'their livelihood is to a large extent dependent on the continued survival of the forest.'[99]

The Tigray state also manages some other forest areas and wildlife reserves, including Hugumburda gra kashu (21,635 ha), Hirmi (30,000 ha), and the Shiraro Kafta Wildlife Reserve (500,000 ha). A total of 380,000 hectares of the Boswellia paprifera forest areas are managed by private and state sectors in Wolquit, Humera, Tahtay Adiabo, Asgede Tembla, and Tselemti. About 40,000 quintals of olibanum and up to 5,000 quintals of gum Arabic are harvested annually from these forests. State- and NGO-supported community-based area enclosures have been implemented to rehabilitate over 270,000 hectares of degraded land since the late 1980s. The local NGO, Relief Society of Tigray (REST), has been a major catalyst in restoring barren landscapes by using area enclosures.[100] Eucalyptus species are widely planted since they grow prodigiously and have a 'relatively high cash value.'[101] Farmers grow eucalyptus around homesteads and on community woodlots but not adjacent to crop fields. In 1997, the Tigray regional government banned the planting of eucalyptus near or on farmlands because of the adverse environmental impacts associated with eucalyptus as well and to preserve agricultural land for food crops.[102] Drought-resistant species, such as *Acacia etbaica, Acacia amythethophylla/senegal*, and *Euclea schimperi*, are also grown in many parts of the state.[103]

The Amhara regional state is among the least forested states in the country. Farmland covers large portions of the state due to thousands of years of human settlement. The state has thirteen regional forest priority areas totalling nearly 600,000 hectares. These forest areas include Woinye (North Wollo), 1,200 hectares; Yegof and Denkoro (South Wollo), 1,800 hectares; Angereb and Western Lowlands (North Gondar), 529,000 hectares; Yeraba and Abafelasse (East Gojam), 1,000 hectares; Sekela Mariam (West Gojjam), 1,100 hectares; Kahtasa and Gwangwa (Awi), 31,000 hectares; Yerke (Oromiya), 6,000 hectares; and Yewof Washa (North Shewa), 17,000 hectares.[104] High forests constitute less than 0.5 per cent of the state's total area, however, while woodland amounts to 4.2 per cent and plantation forests, mostly eucalyptus trees, cover 1.2 per cent.[105]

Only a few pockets of indigenous trees of *Podocarpus* or zigba forests still remain in remote highlands of the South Wollo zone, near the town of Gerba. The regional state has declared sites located on the Kombolcha-Bati road as protected forests. The *Podocarpus* is an evergreen and provides shade and nourishment for small mammals and birds. It is also home to the Colobus Monkey, an animal that is increasingly rare due to the coveted nature of its skin. However, the integrity of these 53 hectares of forests have been somewhat affected by the pines planted along its edges during the tree planting campaigns of the 1980s, though these probably served to protect the forests as well. After a century of heavy harvesting, less than 1 per cent of the original *podocarpus* forests remain in the country. The scarcity of the *podocarpus* tree is partially due to its difficulty reproducing. The tree's oils are also sought after for medicinal purposes.[106]

In the western part of the regional state, the Zeghe Peninsula, which protrudes into Lake Tana, is covered with diverse indigenous forest species, including *Justica schimperiana, Rothmania urcellifomis, Millettia ferruginea, Ehertia cymosa, Celtis* sp., *Croton macrostachyus, Albizia* spp., and *Diospyrossp.*[107]

Small patches of indigenous forests are also found between Ankober and Debre Sina in North Shewa zone, covering over 13,000 hectares of very steep slopes with altitudes ranging between 2,000 metres and 3,700 metres above sea level. The Yewof Washa Forests – as they are called locally – are alpine mixed with broadleaf coniferous forests that include such indigenous species as 'Weira (*Olea europaea* sub species *cuspidata*), Kosso (*Hagenia abyssinica*), Tid (*Juniperus procera*), Zigba (*Podocarpus falcatus*), Imbus (*Allophylus abyssinicus*), Asta (*Erica Arborea*), Amija (*Hypericum revolutum*), and Jibbra (giant *Lobelia* spp.).'[108] About 12 per cent of the endemic plant species in the country are found in these highland forests.[109] Much of these forests are in 'deeply incised valleys near the famous Tekle Haimanot Gedam,' where either the forests have been 'protected by inaccessibility or the moral authority of the Gedam.'[110] Wild fauna in these forests include Gelada Baboon, Colobus Monkey, Anubis Baboon, Vervet Monkey, Wild Cat, Serval, Civet Cat, Leopard, Klipspringer, and Bushbuck.[111] The endemic Ankober Serin makes its home in the Yewof Washa forests.

Because the state is experiencing a severe shortage of wood for fuel, construction, and industries, the regional government is promoting extensive forest development programmes, including the integration of forestry activities (agroforestry) into the farming system. These take the form of alley cropping, planting multipurpose trees on field or farm boundaries, woodlots, grazing lands, hillsides, and in gullies. People also practise multistory cropping where they intercrop trees with an agricultural crop. There has been increased cultivation of trees, especially eucalyptus, around homesteads. It is especially true around towns and along roads. Commercial forestry is being encouraged on 10 hectares or more land leased from the state without fee for 25 years.[112] Farmers seem to be more willing than any time in the past to plant trees on their plots of land because they know that they would be able to garner the benefits of their investment.[113] All these planting activities reduce the rate of deforestation but also control erosion, improve the quality of the soil, generate income, and produce animal feed.

Bamboo forests

Ethiopia is home to the largest area of natural bamboo in Africa; it has about two-thirds of African bamboos and 7 per cent of the world's bamboo resources.[114] It has over a million hectares of bamboo forest, of which about a third is found in Benshangul-Gumuz regional state that borders Sudan, which equals about 6.5 per cent of total forest cover in the country.[115] Ethiopia has two species of bamboo: *Arundinaria alpina* (Alpine or highland bamboo), which grows at between 2,000 metres and 3,200 metres and requires rainfall ranging between 900 mm and 1,400 mm per year; *Oxytenanthera abyssinica* (lowland bamboo) grows at 1,000–1,800 metres – mostly in the northwestern and southwestern lowland areas – and requires a minimum rainfall of 700 mm per year.[116] Approximately 98 per cent of bamboo forest areas are classified as public forest managed by regional states.[117] Bamboo is used for a variety of goods: ceiling, doors, floors, and furniture. It is used as construction material for fences, beehives, mats, baskets, and ropes. It is also used for fuel, but it does not make an excellent fuel as it burns quickly. The bamboo forest can also provide a habitat for diverse life forms. Bamboo is adaptable to a variety of climatic conditions and ecosystems, and vigorous growth makes it a useful plant for carbon sequestration.[118] Planting a fast-growing plant like bamboo on a degraded land can also reduce soil erosion and regenerates a revitalised ecosystem. The government has banned the making of charcoal from burnt wood for merchandising and is vigorously promoting sustainable alternatives such as bamboo.[119]

Fuelwood crisis

In Ethiopia, there is a critical need for woody biomass for fuel. Most Ethiopians depend on fuelwood and charcoal as their primary sources of energy. Modern fuels, except kerosene, are almost entirely unknown in rural areas. At present, more than 95 per cent of the total demand for wood and woody biomass in rural Ethiopia is for fuelwood. In urban areas, fuelwood, charcoal, and dung make up roughly 80 per cent of the total household energy consumption.[120] Overall, biomass accounts for 91 per cent of the country's total energy consumption.[121] Often, people harvest biomass in a sustainable way. They collect dead wood, shrubs, and branches. However, when biomass becomes scarce, they pay less attention to sustainable harvests and deforestation occurs. In most regions of the country, harvesting is well above sustainable yields. Fuelwood scarcity is also having an effect on productivity, as more time must be spent to gather fuel. Firewood scarcity means that women and children, who handle many household chores, must spend hours each day looking for fuel. For those in urban areas, scavenging is not an option; as the costs of fuelwood rise, valuable financial resources must be diverted. The scarcity of fuelwood has been particularly hard on the urban poor, who are forced to spend 20 per cent to 30 per cent of their household income on firewood or charcoal. In the capital of Addis Ababa, this is as high as 50 per cent.[122] Growing rural and urban populations are only exacerbating the problem and placing more pressure on the supply.

As fuelwood supplies have depleted in many of the highland regions, rural people have turned to dried animal manure and crop residues, which often contribute to poor yields as they deprive farms of valuable fertiliser and organic matter.[123] In northern Ethiopia, dung and crop residues account for more than four-fifths of total household energy consumption.[124]

Of the many eucalyptus species that were brought into Ethiopia during the early twentieth century, *Eucalyptus globulus* is the most extensively cultivated species, followed by *Eucalyptus camaldulensis*. For more than one hundred years, *Eucalyptus globulus* has been successfully grown on the fertile, volcanic soils of highland Ethiopia, with annual rainfalls ranging from 800 mm to over 2,000 mm. It is a fast-growing species and 'it can be cut every 2–3 years for over 40 years without being replanted because new sprouts grow from the stump.'[125] The wood burns well and makes good fuelwood and charcoal. Under favourable environmental conditions, *eucalyptus globulus* can easily reach a height of about 60 m and a diameter of 2 m or more and has a 'straight trunk as long as two-third of its total height,'[126] making it useful for telephone and electric poles and construction. It has 'extensive and dense root systems' that allows it to out-compete other plants for 'available soil moisture'[127] and is less susceptible to pests, diseases, and climate variability. The less widely cultivated *Eucalyptus camaldulensis* is also a good fire and charcoal wood and can grow on relatively poor soils and in drier areas.[128] Biomass productivity comparison in Ethiopia's central plateau between the best four eucalyptus species – *E. globulus*, *E. salinga*, *E. grandis*, and *E. vininalis* – and four best exotic conifers and indigenous species indicate that, under most situations, eucalyptus species were found to be the most efficient in converting 'energy and available water into biomass.'[129]

In Ethiopia, eucalyptus species are widely used for reforestation and afforestation because of their fast growth and ability to thrive in relatively poor soils or on degraded land. The area of land covered with eucalyptus trees increased from approximately one million hectares in 1960 to seven million hectares in the late 1980s. Eucalyptus accounts for over two-thirds of the trees planted annually.[130] People value eucalyptus species for a number of reasons. They provide benefits such as fuel and round wood for construction timbers very quickly. They also provide multiple benefits in rural areas – fuel, shade, poles, windbreaks, and soil stabilisation – in a short amount of time.[131] However, eucalyptus species have also been criticised for being a poor species to use in afforestation programmes because of soil and water depletion.[132] Unlike leucaena and acacia, eucalyptus are non-leguminous and do not fix nitrogen into the soil. They, therefore, do not add to the nutrients in the soil and may compete with crops for soil nutrients and moisture if planted adjacent to them. Leguminous trees, by contrast, enhance crop productivity and sustainability.[133] Although there is much evidence that eucalyptus depletes soil nutrients, eucalyptus may also improve some soil conditions. Eucalyptus may help with moisture and soil retention, especially in barren areas.[134]

The history of eucalyptus in Ethiopia spans more than a century. Eucalyptus is now deeply rooted in the Ethiopian landscape, in poor and rich soils, and highland as well as in lowland areas of the country. The widespread planting of eucalyptus has virtually transformed the Ethiopian rural and urban landscape. It is almost impossible to imagine rural as well as urban Ethiopia without eucalyptus growing everywhere. One tends to agree with Jagger and Pender that the environmental impacts of eucalyptus are complex and depend on local conditions. Eucalyptus may have various detrimental effects in moisture-stressed environments, such as northern Ethiopia. Under these circumstances, eucalyptus may have adverse impacts on crops and water resources due to their capacity to compete for light, water, land nutrients, as well as their allelopathic effects. However, there may also be environmental benefits to planting eucalyptus. Eucalyptus can grow rapidly and thrive in such an environment, reduce runoff and control erosion, provide a source of biomass, reduce pressure on natural forests, and survive fires, pests, and diseases. Whether the environmental benefits outweigh the costs depends on the local conditions. Rainfall and availability of water, erosion and runoff risk, the scarcity of land and biomass, and alternative sources of timber and fuel all play a role. Decisions should not be made regarding planting eucalyptus based on isolated negative or positive impacts. The economic implications of these trees on households also need to be considered.[135]

Biomass use will continue to place a strain on Ethiopia's forest resources. Biomass use is unlikely to decrease for a fairly long time because continued population growth in the county would increase overall demand. Also, incomes for the overwhelming majority of the Ethiopian households are unlikely to go up significantly enough to result in substantial substitution of biomass by modern forms of energy.[136] The country possesses a vast amount of natural gas, but the prospect for developing this power shortly remains in doubt owing mainly to political uncertainties in areas where deposits of this energy exist. However, the prospect for hydroelectric production appears promising.

Conclusion

Ethiopia suffers from severe deforestation and environmental degradation. Degradation is a process in which natural resources are depleted or destroyed. One of the most insidious forms is soil erosion. The Ethiopian people have been practising subsistence farming for centuries. Major production systems include crop cultivation and livestock raising based on traditional practices, some of which have led to land mismanagement. Rapid population growth and a shortage of arable land have forced people to clear forests and cultivate hillsides. It has caused land productivity to decline to the point that the land cannot produce enough to feed the human population. This problem is more pronounced in the northern highland regions of the country. There is relatively high forest cover in the southeast and southwest highland regions. However, encroachment, burning, and clearing are stressing these forests. In many areas, understory vegetation is burned or otherwise cleared for the expansion of coffee farms, while, in others, the forest is totally burned or cut to clear space for grain crops.

Legal and illegal exploitation threatens forest resources. A large number of privately or government-owned saw mills exist in areas with relatively high forest cover but a limited resource base. Jimma and Illu Ababor alone have over a dozen sawmills, accounting for nearly one-third of the sawmills in the country, excluding six or so mobile ones. Pit sawing or individual loggers do the majority of illegal harvesting from the forest. They provide forest products like lumber and planks to either individual customers or unauthorised office and household furniture establishments. They are widespread in forested areas as well as agricultural ones. The forest is also exploited for beams for construction, firewood, and charcoal. People increasingly use forests as sources of fuelwood, particularly in enset-root and cereal farming systems. Marketing fuel wood is becoming a major source of income for households and may turn out to be an economic and socio-cultural part of these systems.[137]

The federal government has set up a range of establishments and legislation aimed at administrating, managing, and conserving the nation's forests. However, over time it has become clear that the regional states do not have the capability to manage the nation's forests. Regional state institutions are unable to fulfil their obligations or to implement forest conservation measures. There are many reasons for this, the most important of which stem from a lack of financial, human, and technical resources at all levels. Organisations meant to execute and monitor forest policies are poorly equipped and understaffed. Still, failure to effectively implement forest policies is not only the result of a lack of resources. Institutional and structural malfeasance is also to blame. Structuring and organisational instability have been frequent and ongoing in Ethiopia. There is a lack of decision-making power, information exchange, coordination, and clarity about responsibilities. The separation between the environmental decision-making bodies and those that implement the decisions is also an issue. As a result of these combined factors, state policies on forest use, management, and conservation are ineffectively implemented and do not produce positive results.[138]

Not only are state forest policies improperly implemented but also the policies themselves are often exclusionary. Despite the institutionalisation of decentralised governance, top-down decision-making still exists in state forest policies. There is a lack of cooperation between the state and communities that use the forests. State forest policies view local forest users as intruders rather than part of the ecosystem who should be included in forest management and conservation. Such exclusionary policies simply force local forest users to pursue illegal activities. The new idea of participatory forest management (PFM), which is aimed at making local forest users the centre of the management and conservation strategies, is a prudent step. However, future PFM projects should be built into existing traditional institutions and community-initiated local structures instead of trying to create entirely new institutional structures that have often failed to achieve their aims in the past.

Stellmacher's detailed study of coffee forests in Kaffa and the Bale Mountains clearly illustrates that conservation efforts should translate traditional forest-use arrangement into official forest management policy. In other words, the local system in which local peasants have usufruct rights to plots of forest should be backed up by official policy. Traditional forest uses should be legal. Local communities should have the right to use forest plots for NTFPs, which is necessary for income generation. The state imposes restrictions on indigenous households, which have a vested interest in conserving forests from NTFPs. Also included should be collective use of other forest resources, such as wood, for the whole community. The same rationale holds true for local mechanisms meant to enforce these arrangements. Instead of establishing new entities such as the Forest User Society (FUS), existing structures should be incorporated into conservation measures and strengthened.[139]

There is also a need for regional government capacity building. The decentralisation of power from federal to regional states to carry out development and conservation plans suggests that regional states have the ability to undertake such responsibilities. But the fact of the matter is that not all regions have adequate capabilities to execute development and conservation programmes. Few regions have existing technical capacities, but vast deficiencies exist everywhere. Furthermore, the government must train professional personnel who can guide conservation and development projects in communities across regional states.

While the government needs to do a whole lot to address the country's critical environmental problems, commitment to and support for the conservation of the country's natural resources at the federal government level is not lacking, despite policy shortcomings mentioned previously. The federal government invests a substantial amount of resources in numerous conservation and environmental rehabilitation projects around the country. It also funds several environmental research organisations. Two federally funded organisations manage the country's plant and animal genetic resources: the Ethiopian Plant Genetic Resources Center (PGRC/E) and the Institute of Biodiversity Conservation and Research (IBC). Established in 1976, the PGRC/E collects, evaluates, documents, conserves, and promotes the use of crop germplasm occurring in the country. It currently holds over 60,000 accessions of over 100 plant species that are appropriately preserved. Cereals and pulses constitute the majority of collected species. The IBC, which was created in 1998, conducts biodiversity research and research in advancing the development and sustainable utilisation of the country's plant, animal, and microbial genetic resources.[140] Several other government-funded research organisations are also involved in soil and water conservation research endeavours and pilot projects. The maintenance of biodiversity is crucial for the country for a number of reasons, 'including the formation of soils, the production of food, the purification of water, the breakdown or decomposition of waste products, the maintenance of the composition of gases in the atmosphere, and other' essential 'functions that contribute to the fundamental stability of ecosystems.'[141]

Notes

1 Gebre Markos Wolde Selassie and Deribe Gurmu. 2001. Problem of forestry associated with institutional arrangements. In: *Imperative Problems Associated with Forestry in Ethiopia: Proceedings of a Workshop, Biological Society of Ethiopia*, February 1, 2001, Addis Ababa, pp. 45–80; p. 48.

2 Stanislaw Chojnacki. 1963. Forests and the forestry problems as seen by some travelers in Ethiopia. *Journal of Ethiopian Studies* 1, 1, pp. 32–9; Girma Kebbede, *The State and Development in Ethiopia*, Atlantic Highlands, NJ: Humanities Press, 1992, pp. 57–8.

3 S. Bristow. 1995. *Agroforestry and Community-Based Forestry in Eritrea*. Working Paper No. 2. Washington, DC: World Bank; J. Bruce, A. Hoben, and D. Rahmato. 1994. *After the Derg: An Assessment of Rural Land Tenure Issues in Ethiopia*, Madison: Land Tenure Center, University of Wisconsin at Madison; Pamela Jagger and John Pender. 2000. *The Role of Trees for Sustainable Management of Less-Favored Lands: The Case of Eucalyptus in Ethiopia*, Washington, DC: International Food Policy Research Institute, Environment and Production Technology Division, EPTD Discussion Paper No. 65, Table 3, p. 8. Available online at http://ageconsearch.umn.edu/bitstream/16122/1/ep000065.pdf (accessed 21 April 2011).

4 Iain Darbyshire and Mohammed Umer. 2003. Forest clearance and regrowth in northern Ethiopia during the last 3000 years. *The Holocene* 13, 4: 537–46; p. 1.

5 Ibid.

6 Daniel Gamachu. 1988. *Environment and Development in Ethiopia*, Geneva: International Institute for Relief and Development.

7 EWNHS, p. 59.

8 Richard Pankhurst. 1962. The foundation and growth of Addis Ababa to 1935. *Ethiopian Observer* 6: 49–51; Ronald J. Horvath. 1968. Addis Ababa's eucalyptus forest. *Journal of Ethiopian Studies* 6, 1: 13–9.

9 Richard St. Barbe Baker. 1964. Some reflections on trees and forests for Ethiopia. *Ethiopia Observer* 8, 2: 189–92; p. 190.

10 Richard Pankhurst. 1992. The history of deforestation and afforestation in Ethiopia prior to World War II. *Ethiopian Institute of Development Research* 2, 2: 59–77.

11 W.E. Logan. 1946. *An Introduction to the Forest of Central and Southern Ethiopia*, Oxford: Imperial Forestry Institute and Oxford University Press, Institute Paper No. 224, p. 44.

12 G. Conn. 1991. *Forest Management in Ethiopia: ca 1600 to Present*. A Contribution to the Task Force on Forest Management within the Working Group Conservation and Development of Forest Resources, Addis Ababa, Ministry of Agriculture.

13 Ibid.

14 Yonas Yemshaw. 2001. Status and prospects of forest policy in Ethiopia. In: *Imperative Problems Associated with Forestry in Ethiopia, Proceedings of a Workshop*. Biological Society of Ethiopia, February 1, Addis Ababa, p. 16.

15 Melesse Damte. 2001. Land use and forest legislation for conservation, development and utilization of forests. In: *Imperative Problems Associated with Forestry in Ethiopia: Proceedings of a Workshop*. Biological Society of Ethiopia, February 1, Addis Ababa, pp. 34–5.

16 Georgi Galperin. 1981. *Ethiopia: Population, Resources, and Economy*, Moscow: Progress Publishers, p. 231; Conn, *Forest Management in Ethiopia: ca 1600 to Present*.

17 Conn, *Forest Management in Ethiopia: ca 1600 to Present*.

18 H. P. Haffnagel. 1961. *Agriculture in Ethiopia*, Rome: FAO, p. 420.

19 Till Stellmacher. 2006. *Governing the Ethiopian Coffee Forests: A Local Level Institutional Analysis in Kaffa and Bale Mountains*. PhD dissertation, Institutfür Lebensmittel- und Ressourcenökonomik (ILR), Bonn, p. 17.

20 Ibid.

21 Melesse Damte, p. 35.

22 Girma Amente. 2005. *Rehabilitation and Sustainable Use of Degraded Community Forests in the Bale Mountains of Ethiopia*. PhD dissertation, Faculty of Forest and Environmental Sciences, Albert-Ludwigs-University, Freiburg im Breisgau, Germany, p. 9.

23 Stellmacher, p. 17.

24 Ermias Bekele. 1989. *Inventory of forestry projects in Ethiopia*, Consultant to the World Bank, Addis Ababa, Unpublished manuscript; Y. Yemshaw. 2002. Overview of forest policy and strategy issues in Ethiopia. In: *Forests and Environment: Proceedings of the Fourth Annual Conference*, 14–15 January, Forestry Society of Ethiopia, Demel Teketay and Yonas Yemshaw (eds.), Addis Ababa, Ethiopia; K. Mengistu. 2003. Ethiopia country paper. In: *Proceedings of Workshop on Tropical Secondary Forest Management in Africa: Reality and Perspectives*, 9–13 December, Nairobi, Kenya, Rome: FAO.

25 John Campbell. 1991. Land or peasants? The dilemma confronting Ethiopian resource conservation. *African Affairs* 90: 5–21; p. 12.

26 Kurt Janssen, Michael Harris, and Angela Penrose. 1987. *The Ethiopian Famine*, London: Zed Press, p. 121; Daniel Gamachu, p. 22; Girma Kebbede, pp. 87–9.

27 Daniel Gamachu, p. 23.

28 Jonathan McKee. 2007. *Ethiopia: Country Environmental Profile, 2007*, p. 28. Available online at http://ec.europa.eu/development/icenter/repository/Ethiopia-ENVIRONMENTAL-PROFILE-08-2007_en.pdf (accessed 5 April 2011).

29 Kebbede, pp. 88–9.

30 Farm-Africa. 1996. *Bonga Forest Conservation Project Proposal*, Addis Ababa: Farm-Africa, p. 5.

31 Siegfried Pausewang. 2002. No environmental protection without local democracy. In: *Ethiopia: The Challenge of Democracy From Below*, Bahru Zewde and Siegfried Pausewang (eds.), Addis Ababa: Forum for Social Studies, pp. 87–100; pp. 93–5.

32 Ibid.

33 Melesse Damte, p. 36.

34 Ethiopian Forest Action Program (EFAP) 1994. *Ethiopian Forestry Action Program*, Addis Ababa: EFAP.

35 Stellmacher, p. 19; BCEOM-French Engineering Consultant. 1989. *Abbay River Basin Integrated Development*. Mater Plan, Phase 2: Data Collection, Site Investigation Survey Analysis, Section II: Sectoral Studies, Volume 1, pp. 3–18. In 2007, most of these NFPAs were transferred to the management bureaus of the different regional governments and are now referred to as Regional Forest Priority Areas (RFPAs). See Alemu Mekonnen and Randall Bluffstone. 2008. Policies to increase forest cover in Ethiopia: lessons from economics and international experience. In: *Proceedings of Workshop: Policies to Increase Forest Cover in Ethiopia*, Sisay Nune, Alemu Mekonnen, and Randall Bluffstone (eds.), 2008. Addis Ababa: Ministry of Agriculture and Rural Development, Ethiopia, pp. 23–69; p. 23.

36 Melesse Damte, p. 39.

37 Woody Biomass Inventory and Strategic Planning Project (WBISPP). 2004. *Woody Biomass Yield Study: Incremental Stem Monitoring and Wood Ring Analysis*, Addis Ababa: WBISPP.

38 EPA, State of Environment Report for Ethiopia, 2003; Cited in Jonathan McKee, *Ethiopia: Country Environmental Profile, 2007*, p. 48. Available online at http://ec.europa.eu/development/icenter/repository/Ethiopia-ENVIRONMENTAL-PROFILE-08-2007_en.pdf (accessed 5 April 2011).

39 Alemu Mekonnen and Randall Bluffstone. 2008. Policies to increase forest cover in Ethiopia: lessons from economics and international experience. *Proceedings of Workshop: Policies to Increase Forest Cover in Ethiopia*. Sisay Nune, Alemu Mekonnen, and Randall Bluffstone (eds.), Addis Ababa: Ministry of Agriculture and Rural Development, Ethiopia, p. 34; M. Reusing. 1998. *Monitoring of Forest Resources in Ethiopia*. Government of the Federal Democratic Republic of Ethiopia, Ministry of Agriculture, Natural Resources Management & Regulatory Department & German Agency for Technical Cooperation (GTZ), Addis Ababa.

40 Melesse Damte, pp. 24–37.

41 Alemu Mekonnen and Randall Bluffstone, p. 39.

42 McKee, p. 29.

43 Markos Ezra, 1997. *Demographic Response to Ecological Degradation and Food Security: Drought Prone Areas in Northern Ethiopia*, Bloomington: Purdue University Press, pp. 62–105.

44 Sisay Nune. 2008. *Flora Biodiversity Assessment in Bonga, Boginda and Mankira Forest, Kaffa, Ethiopia*, Addis Ababa. Available online at www.kafa-biosphere.com/assets/content-documents/KafaFloral-Survey-Final-Report.pdf (accessed 4 April 2011).

45 Kidane Mengistu. 2002. *Ethiopia Country Paper: Workshop on Tropical Secondary Forest Management in Africa: Reality and Perspectives*. Organised in collaboration with ICRAF and CIFOR Nairobi, Kenya, 9–13 December. Available online at www.fao.org/docrep/006/j0628e/J0628E50.htm (accessed 15 April 2011).

46 Tsegaye Fikadu. 2008. Overview of natural resources in SNNPR. In: *Proceedings of Workshop: Policies to Increase Forest Cover in Ethiopia*, Sisay Nune, Alemu Mekonnen, and Randall Bluffstone (eds.), Addis Ababa: Ministry of Agriculture and Rural Development, Ethiopia, pp. 92–103; p. 95.

47 Nigussu Feyissa. 2008. Forest resources of Oromiya National Regional State. In: *Proceedings of Workshop: Policies to Increase Forest Cover in Ethiopia*, Sisay Nune, Alemu Mekonnen, and Randall Bluffstone (eds.), Addis Ababa: Ministry of Agriculture and Rural Development, Ethiopia, pp. 83–91; pp. 84–8.

48 Ibid., pp. 83–91; p. 88.

49 Farm-Africa. 1996. *East Langano Development and Conservation Project*, Addis Ababa: Farm-Africa. Oromiya Regional State. 1996. *The Resource Base, Its Utilization and Planning for Sustainability*. Regional Conservation Strategy Task Force, Addis Ababa, p. 42.

50 Oromiya Regional State, p. 42.

51 EWNHS. 1996. *Important Bird Areas of Ethiopia: A First Inventory*, Addis Ababa: EWNHS, pp. 183–4.

52 Ministry of Water Resources (MoWR). 1996. *Omo-Gibe River Basin Integrated Development Master Plan Study*. Final Report, Volume III, Development Zones and Areas; Oromiya Regional State, p. 42.

53 Kidane Mengistu. 2002. *Ethiopia Country Paper: Workshop on Tropical Secondary Forest Management in Africa: Reality and Perspectives*. In collaboration with ICRAF and CIFOR Nairobi, Kenya, 9–13 December. Available online at www.fao.org/docrep/006/j0628e/J0628E50.html (accessed 9 July 2016).

54 UNESCO. 2010. UNESCO announces selection of 13 New Biosphere Reserves. Available online at www.unesco.org/new/en/media-services/single-view/news/unesco_announces_selection_of_13_new_biosphere_reserves/#.V30vYeeAOko (accessed 6 July 2016).

55 Institute of Biodiversity Conservation (IBC). 2008. *Ecosystems of Ethiopia*, Addis Ababa: IBC.

56 Kidane Mengistu, p. 14.

57 BirdLife International. 2011. *Sites-Important Bird and Biodiversity Areas: Menagesha State Forest*. ET031. Available online at www.birdlife.org/datazone/sitefactsheet. php?id=6265 (accessed 6 July 2016).

58 EWNHS, p. 173.

59 EWNHS, pp. 59–62.

60 Kristin Underwood. 2009. Eliminating eucalyptus to aid Ethiopia, *Travel & Nature*, 15 September. Available online at www.treehugger.com/files/2009/09/eliminating-eucalyptus-to-aid-ethiopia.php (accessed 21 April 2011); Available online at http://ewnhs.org.et/wp-content/uploads/downloads/2011/03/Biodiversity-Hotspots-of-Ethiopia.pdf (accessed 4 November 2014).

61 EWNHS, A Glimpse at Biodiversity Hotspots of Ethiopia. Available online at http://ewnhs.org.et/wp-content/uploads/downloads/2011/03/Biodiversity-Hotspots-of-Ethiopia.pdf (accessed 9 July 2016).

62 Stellmacher, p. 30.

63 Stellmacher, p. 30.

64 IBC, *Ecosystems of Ethiopia*.

65 Melesse Damte, pp. 23-25; Kidane Mengistu, p. 16.

66 Ethiopian Wildlife Conservation Authority (EWCA). 2007. *Bale Mountains National Park: General Management Plan 2007, 2017*, Addis Ababa: EWCA, p. 38.

67 Kidane Mengistu, p. 17.

68 Stellmacher, p. 31.

69 Stellmacher, p. 31; EWCA, p. 41.

70 C. W. George and R. W. Mutch. 2001. *Ethiopia: Strengthening Forest Fire Management*. FAO Project Document (TCP/ETH/0065), April, Rome; Yonas Yemshaw, p. 14.

71 Dechassa Lemessa and Matthew Perault. 2011. *Forest Fires in Ethiopia: Reflections on Socio-Economic and Environmental Effects of the Fires in 2000*. Available online at http://reliefweb.int/report/ethiopia/forest-fires-ethiopia-reflections-socio-economic-and-environmental-effects-fires (accessed 6 July 2016).

72 EWCA, pp. 34–6.

73 Celia A. Harvey and David Pimentel. 1996. Effects of soil and wood depletion on biodiversity. *Biodiversity and Conservation* 5: 1121–30; p. 1125.

74 EWNHS, p. 131.

75 EWNHS, p. 132; BirdLife International. 2016. Sites-Important Bird and Biodiversity Areas (IBAs): *Anferara Forests*. ET058. Available online at www.birdlife.org/datazone/sitefactsheet.php?id=6292 (accessed 6 July 2016).

76 Alemu Mekonnen and Randall Bluffstone, p. 27.

77 Tsegaye Fikadu. 2008. Overview of natural resources in SNNPR. In: *Proceedings of workshop: Policies to Increase Forest Cover in Ethiopia*, Sisay Nune, Alemu Mekonnen, and Randall Bluffstone (eds.), Addis Ababa: Ministry of Agriculture and Rural Development, Ethiopia, pp. 92–103; p. 94.

78 Environment News Service (ENS). 2010. *UNESCO Approves 13 New Reserves for Enhanced Protection*. Available online at www.ens-newswire.com/ens/jun2010/2010-06-07-02.html (Accessed 6 July 2016).

79 Ensermu Kelbessa and Teshome Soromessa. 2004. *Biodiversity, Ecological and Regeneration Studies in Bonga, Borena and Chilimo Forests*. Technical Report Prepared for FARM-Africa, SOS-Sahel, Addis Ababa; C. B. Schmitt. 2006. *Montane Rainforest with Wild Coffea Arabica in the Bonga Region (SW Ethiopia): Plant Diversity, Wild Coffee Management and Implications For Conservation*. Ecology and Development Series No. 47, Cuvillier Verlag Göttingen. Cited in Sisay Nune. 2008. *Flora Biodiversity Assessment in Bonga, Boginda and Mankira Forest, Kaffa, Ethiopia*, Addis Ababa. Available online at www.kafa-biosphere.com/assets/content-documents/KafaFloral-Survey-Final-Report.pdf (accessed 4 April 2011).

80 B. Tola and O. Wirtu. 2004. *Marketing Study of NTFPs in the South West, NTFP Project.* Cited in Jonathan McKee. 2007. *Ethiopia: Country Environmental Profile 2007*, p. 43; Yihenew Zewdie. 2003. *Forest access: policy and reality in Kafa, Ethiopia.* Available online at www.agriculturesnetwork.org/magazines/global/rights-and-resources/forest-access-policy-and-reality-in-kafa-ethiopia (accessed 6 July 2016).

81 Stellmacher, p. 23.

82 Taye Bekele. 2003. *Potential of Bonga Forest for Certification: A Case Study Institute of Biodiversity Conservation and Research (IBCR)*, Addis Ababa, Ethiopia. Available online at www.researchgate.net/publications/242305564_THE_POTENTIAL_OF_BONGA_FOREST_FOR_CERTIFICATION_A_CASE_Study (accessed 9 July 2016); SisayNune. 2008. *Flora Biodiversity Assessment in Bonga, Boginda and Mankira Forest, Kaffa, Ethiopia*, Addis Ababa. Available online at www.kafa-biosphere.com/assets/content-documents/KafaFloral-Survey-Final-Report.pdf (accessed 4 April 2011).

83 Dennis Riechmann. 2007. *Literature Survey on Biological Data and Research Carried Out in Bonga Area, Kaffa, Ethiopia.* Available online at www.kafa-biosphere.com/assets/content-documents/Literature-Survey-for-BR-Kafa.pdf (accessed 9 April 2011).

84 Tsegaye Fikadu, pp. 94–5.

85 Adrian P. Wood. 1993. Natural resource conflicts in south-west Ethiopia: state, communities, and the role of the national conservation strategy in the search for sustainable development. *Nordic Journal of African Studies* 2, 2: 83–99; pp. 88–9; Sisay Nune. 2008. *Flora Biodiversity Assessment in Bonga, Boginda and Mankira Forest, Kaffa, Ethiopia*, Addis Ababa. Available online at www.kafa-biosphere.com/assets/content-documents/KafaFloral-Survey-Final-Report.pdf (accessed 4 April 2011).

86 Sisay Nune, p. 39.

87 Girma Balcha, Kumelachew Yeshitela, and Taye Bekele (eds.). 2004. *Proceedings of a National Conference on Forest Resources of Ethiopia: Status, Challenges and Opportunities*, 27–29 November 2002, Addis Ababa; Ensermu Kelbessa and Teshome Soromessa. 2004. Biodiversity, *Ecological and Regeneration Studies in Bonga, Borena and Chilimo Forests.* Technical Report Prepared for FARM-Africa, SOS-Sahel, Addis Ababa; Sisay Nune, p. 41.

88 Stellmacher, pp. 24–5.

89 Kidane Mengistu, p. 19.

90 Tsegaye Fikadu, p. 98.

91 Omer Redi. 2010. Ethiopia – They Have Become Farmers of Trees, *African Voices.* Available online at http://it-it.facebook.com/note.php?note_id=434666237608 (accessed 2 April 2011). Over a third of household's income is derived from the sale of these products. Yihenew Zewdie. 2003. *Forest Access: Policy and Reality in Kaffa, Ethiopia*, Agri-Cultures, September. Available online at www.agriculturesnetwork.org/magazines/global/rights-and-resources/forest-access-policy-and-reality-in-kafa-ethiopia (accessed 1 April 2011).

92 UNESCO. 2010. *Kaffa Biosphere Reserve Recognized as UNESCO Biosphere Reserve.* Available online at www.kafa-biosphere.com/ (accessed 6 April 2011).

93 Omer Redi. 2010. Ethiopia – They Have Become Farmers of Trees, African Voices, 20 November 2010. Available online at http://it-it.facebook.com/note.php?note_id=434666237608 (accessed 2 April 2011). Farm Africa has also been assisting Participatory Forest Management (PFM) efforts in Kaffa since 1996. Farm-Africa.1996. *Bonga Forest Conservation Project Proposal*, Addis Ababa: Farm-Africa; Tsegaye Fikadu, p. 99; Southern Nations, Nationalities and Peoples' Regional State. 2005. *Conservation Strategy of the Region, Environmental Protection, Land Administration and Use*, Hawassa Ethiopia.

94 Tsegaye Fikadu, pp. 94–5.

95 Kidane Mengistu, p. 23.
96 United Nations Educational, Scientific, and Cultural Organization, World Water Assessment Program. 2004. *National Water Development Report for Ethiopia, Final Report*, Addis Ababa, Ministry of Water Resources, December 2004, UN-WATER/WWAP/2006/7, p. 144. Available online at http://unesdoc.unesco.org/images/0014/001459/145926e.pdf (accessed 26 November 2010).
97 Tsegaye Fikadu, pp. 94–5.
98 Yearswork Admassie. 1997. *Forest, Animals and People: Survival through Harmony, A Study of the Desse'a Forest in Eastern Tigray*. Paper presented at a Conference on Environment and Development in Ethiopia, held in Debre Zeit, 12–15 June, p. 12.
99 Ibid.
100 Nigus Esmaile. 2008. Forest resources of Tigray National Regional State. In: *Proceedings of Workshop: Policies to Increase Forest Cover in Ethiopia*, Sisay Nune, Alemu Mekonnen, and Randall Bluffstone (eds.), Addis Ababa: Ministry of Agriculture and Rural Development, Ethiopia: 77–81; pp. 78–80.
101 Ibid.
102 Pamela Jagger and John Pender. 2000. *The Role of Trees for Sustainable Management of Less-Favored Lands: The Case of Eucalyptus in Ethiopia*. Washington, DC: International Food Policy Research Institute, EPTD Discussion Paper No. 65.
103 Relief Society of Tigray (REST). 1995. *Farming Systems, Resource Management and Household Coping Strategies in Northern Ethiopia, Report of a Social and Agro-Ecological Baseline Study in Central Tigray*. Prepared by Gebremedhin Gebru of the REST and Dr. Arne Olav Oyhus of NORAGRIC at the Agricultural University of Norway, no place of publication, p. 32.
104 Håkan Sjöholm. 2008. A review of some forest management experiences from Ethiopia and Tanzania. In: *Proceedings of workshop: Policies to Increase Forest Cover in Ethiopia*, Sisay Nune, Alemu Mekonnen, and Randall Bluffstone (eds.), pp. 149–72; p. 161.
105 Sisay Nune, Alemu Mekonnen, and Randall Bluffstone, p. 73.
106 Alula Pankhurst. 2000. Environment: Awliyaw: the largest and oldest tree in Ethiopia? *Addis Tribune* 18 February.
107 Available online at http://ewnhs.org.et/wp-content/uploads/downloads/2011/03/Biodiversity-Hotspots-of-Ethiopia.pdf (accessed 4 November 2014).
108 EWNHS, p. 81.
109 Available online at http://ewnhs.org.et/wp-content/uploads/downloads/2011/03/Biodiversity-Hotspots-of-Ethiopia.pdf (accessed 4 November 2014).
110 Farm-Africa. 1996. *Mitak Natural Resources Conservation Project, North Shewa, Amhara Regional State*, Addis Ababa, Community Forest and Wildlife Conservation Project, p. 5.
111 Available online at http://ewnhs.org.et/wp-content/uploads/downloads/2011/03/Biodiversity-Hotspots-of-Ethiopia.pdf (accessed 4 November 2014).
112 Woreta Abera, Amhara National Regional State's (ANRS). 2008. Efforts Towards Forest Cover Increment. In: *Proceedings of workshop: Policies to increase forest cover in Ethiopia*, Sisay Nune, Alemu Mekonnen, and Randall Bluffstone, pp. 73–6; pp. 73–4.
113 Farm-Africa, p. 9.
114 Kathleen Buckingham. 2014. Rebranding bamboo for Bonn: the 5 million hectare restoration pledge, *World Resources Institute*, 23 December 2014. Available online at www.academia.edu/9877484/Rebranding_Bamboo_for_Bonn_The_5_Million_Hectare_Restoration_Pledge (accessed on 5 March 2016).
115 International Network for Bamboo and Rattan (INBAR). 2010. Study on utilization of lowland bamboo in Benshangul-Gumuz region, Ethiopia. Available online at

www.cangoethiopia.org/assets/docs/Study%20on%20Utilization%20of%20 Bamboo%20in%20BG%20Region.pdf (accessed 5 April 2014).

116 Ibid.; The total bamboo area in Ethiopia is 1 million ha (source: Zenebe Mekonnen, Adefirse Worku, Temesgen Yohannes, Mehari Alebachew, Demel Teketay, and Habtemariam Kassa. 2014. Bamboo resources in Ethiopia: Their value chain and contribution to livelihoods. Available online at journals.sfu.ca/era/index.php/era/article/view/923/624 (accessed 9 July 2016).

117 FAO, Extent and characteristics of bamboo resources. Available online at ftp://.fao.org/docrep/fao/010/a1243e/a1243e03.pdf, p. 19 (accessed 5 April 2014).

118 Ibid.

119 Ed McKenna. Ethiopia Leads the Bamboo Revolution, Inter Press Service (IPS). Available online at www.ipsnews.net/2013/04/expanding-ethiopias-bamboo-sector/ (accessed 5 April 2014).

120 Woldamlak Bewket. 2005. Biofuel Consumption, Household Level Tree Planting and Its Implications for Environmental Management in the Northwestern Highlands of Ethiopia, *EASSRR* 21, 1: 19–38.

121 Mebratu, D. and M. Tamire. 2002. Energy in Ethiopia: Status, Challenges and Prospects, In: *Proceedings of the Energy Conference*, Addis Ababa, Ethiopia.

122 Kebbede, p. 73.

123 Kebbede, p. 72.

124 A. Bekele-Tesemma. 1997. *A participatory agroforestry approach for soil and water conservation in Ethiopia*. Tropical Resource Management Papers No. 17. Wageningen: Wageningen Agricultural University. Moreover, animal manure and agricultural wastes are used very inefficiently. The efficiency in the use of fuelwood in the traditional way varies from 5 per cent to 10 per cent, indicating a loss of 90–95 percent of total caloric value of the fuel. The traditional conversion of fuelwood into charcoal is also carried out with very low efficiency of about 15 per cent. See Hailu Sharew. 1983. Fuelwood and energy development for African women: implications for Ethiopia. In: *Women in Agricultural Development*, National Workshop, Hawassa, Ethiopia, 26 July–2 August, p. 73.

125 National Academy of Sciences. 1980. *Firewood Crops: Shrub and Tree Species for Energy production*, Washington, DC: National Academy of Sciences, pp. 179–81.

126 Ibid., p. 182.

127 M.E.D. Poore and C. Fries (FAO). 1985. *The Ecological Effects of Eucalyptus*, FAO Forest Paper 59, Rome: FAO, p. 21.

128 National Academy of Sciences, p. 131.

129 Pamela Jagger and John Pender. 2000. *The Role of trees for Sustainable Management of Less-Favored Lands: The Case of Eucalyptus in Ethiopia*, Washington, DC: International Food Policy Research Institute, Environment and Production Technology Division, EPTD Discussion Paper No. 65, Table 3, p. 9. Available online at http://age-consearch.umn.edu/bitstream/16122/1/ep000065.pdf (accessed 21 April 2011).

130 John Davidson. 1989. *The Eucalyptus Dilemma: Arguments For and Against Eucalyptus Planting in Ethiopia*, The Forest Research Center, Seminar Note No. 1.

131 Ibid.

132 A. Michelsen, N. Lisanework, and I. Friis. 1993. Impacts of tree plantations in the Ethiopian highland on soil fertility, shoot and root growth, nutrient utilization, and mycorrhizal colonization. *Forest Ecology and Management* 61: 299–324.

133 Jagger and Pender, pp. 5–6.

134 Ibid., p. 8.

135 Ibid.
136 Elizabeth Cushion, Adrian Whiteman, Gerhard Dieterle. 2010. *Bioenergy Development: Issues and Impacts for Poverty and Natural Resource Management.* Washington, DC: The World Bank, p. 123.
137 WBISPP, p. 155.
138 Stellmacher, pp. 34–5.
139 Ibid., p. 39.
140 McKee, p. 54.
141 Sharon L. Spray and Karen L. McGlothlin (eds.). 2003. *Exploring Environmental Challenges: Loss of Biodiversity*, New York: Rowan and Littlefield Publishers, p. xvi.

5   Freshwater

Ethiopia is a major source of freshwater for northeastern Africa. Rising high above the surrounding areas, it supplies its neighbours with both freshwater and vast amounts of alluvial soil through rivers that originate from its territory. Many of its rivers are short mountain streams. At their upper reaches, they run through deep and narrow canyons and are strongly influenced by rainfall. Many form numerous rapids and waterfalls. As a result, their use is limited because they are not navigable, and only a few of them have broad valleys to pursue irrigation agriculture. Utility is usually higher at their lower ends, but this part of their course often falls outside Ethiopia. Still, freshwater represents Ethiopia's most plentiful natural resource. Overall, annual rainfall is adequate, and the country has several major lakes and rivers and significant groundwater resources. Ethiopia's per capita share of a renewable water resource is better than most Sub-Sahara African countries. However, this situation is changing with increasing water demand induced by high population growth and increased socio-economic development activities. This chapter deals with the state and distribution of freshwater resources in the country, threats to water ecosystems, and the challenges of water resource development and management.

The state and distribution of freshwater resources

Eight major river systems traverse the country's extensive surface: the Blue Nile (locally known as Abbay), Awash, Baro, Dawa, Genale, Omo, Wabi Shebelle, and Tekeze (see Figure 5.1). There are several major lakes of varying depth and sizes. Every year an estimated 1.3 trillion cubic metres (m^3) of rainwater is discharged into these water sources.[1] The total surface water potential of the country is estimated to surpass 110 billion m^3 annually,[2] half of which comes from the Blue Nile System, 14 per cent through the Omo River in the south, and 12 per cent through the Baro-Akobo system in the southwest.[3] Though Ethiopia possesses large quantities of freshwater, uneven distribution of water and human settlements pose significant obstacles for freshwater use and accessibility.

While groundwater is a vital source of water for domestic purposes, industries, and livestock, it is not found in plentiful supplies due to the impermeability of the crystalline rocks.[4] Estimates of potential safe yields of groundwater vary. The Ministry of Water Resources estimates the total yield at about 2.6 billion m^3 per year.[5] But other estimates are more generous. The Ethiopian Institute of Geological Surveys puts the estimate at 6.5 billion m^3. The availability of groundwater is largely dependent on geology. The complex geologic history of Ethiopia has created an uneven distribution of these water resources.[6] The highland areas that experience precipitation levels of 2,000 to 2,400 millimetres (mm)/year have the lowest infiltration capacity due to their extensive metamorphic bedrock or crystalline basement, which has almost zero primary porosity.[7] The lowland areas are the most arid, sometimes receiving only 200 mm of rain per year or less, yet they have the highest infiltration capacities as they overlie sandstones, gravel, and alluvial deposits. Apart from the inverse relationship between infiltration rates and precipitation, there is also a detrimental relationship between topography and rainfall. The highlands have steep slopes, rocky escarpments, and narrow gorges and are also currently experiencing significant erosion due to deforestation and overproduction by farmers. Water is driven by gravity but also is not stopped or slowed by vegetation or soils. Consequently, recharge rates in the highlands are too low for a mass exploitation of groundwater deposits.[8] In the highlands, groundwater recharge is eight to 20 per cent of precipitation due to the high level of rainfall. In the lowlands, recharge rates drop to below five per cent even as infiltration rates rise.[9]

In the lowlands, precipitation is absorbed into the porous ground almost instantaneously, creating an ideal aquifer environment. But there is so little rainfall that the groundwater yields are insufficient for more than livestock watering and domestic water supplies.[10] The primary source of groundwater is the subsurface recharge and runoff from the highlands, with recharge rates declining with distance into the lowlands.[11] Aquifers across Ethiopia all have low-to-moderate productivity due to either low infiltration in the highlands or little recharge in the lowlands. Over 4,000 boreholes tap into the country's groundwater year-round.[12]

Because many groundwater locations are 'highly mineralized and nonpotable,'[13] only a small percentage of Ethiopia's groundwater is usable. Across the country there are issues with unacceptably high and low levels of nitrates, arsenic, iron and manganese, and iodine.[14] In the highlands, groundwater frequently contains sodium, calcium, and magnesium bicarbonates. In the Rift Valley, these bicarbonates also occur, along with sulphate, chloride, and fluoride classified as sodium bicarbonate type. The groundwater in the Rift Valley has extremely high fluoride and salt levels. This poses a difficulty for households in the Rift Valley that depend on the untreated groundwater. Apart from detrimental health effects, the taste of water may also drive water users to more dangerous surface water sources.[15]

Ethiopian groundwater is minimally affected by anthropogenic pollution. However, there are localised exceptions. Nitrate levels in the water table tend to be dangerously high in several urban areas, especially Addis Ababa and Dire Dawa. Leaky septic tanks primarily cause high nitrate rates and are exacerbated when the water table is close to the surface. Other causes of high nitrate levels are the evaporation of saline groundwater and fertiliser runoff from irrigated land.[16]

Ethiopia possesses a vast amount of thermal groundwater. Much of it is in the Rift Valley in volcanically active zones that cover about 100,000 square kilometres (km^2) of the surface.[17] Potential areas of thermal groundwater include the Dallol, Tendaho, and Aluto areas in the northern part of the Rift Valley and the lake region in the middle Rift Valley. Active hot springs are plentiful in the Rift Valley, including the shores of lakes Shala and Langano and in Wondo Genet, Aluto, Boku, Sodere, Gidabo, and Beseka. Higher altitude areas, such as the Addis Ababa Filwuha, Ambo, and Woliso districts, have thermal springs.[18]

River basins

The most important river in Ethiopia is the Blue Nile or Abbay, as locally called. The Blue Nile originates as a small stream near Mount Denguiza in West Gojjam, flows north into Lake Tana, exits from the southeastern corner of the lake, and then takes a wide semicircle to the east and drops like a waterfall at Tis Isat, 30 kilometres south of Lake Tana. From there it flows south and west through deep canyons between Mount Choke in East Gojjam and Mount Amba Farit in South Wollo. On its 900-kilometres course within Ethiopia, several tributaries, including the Beshillo, Jamma, Muger, Guder, Fincha'a, Didessa, and Dabus join it. These tributaries are all on the left-hand side, and like the major river, are perennial streams. The right-hand tributaries, including the Bolassa, Habad, and the Dinder, are steep and torrential.[19] Upon reaching the Senna plains, it flows north into the clay flatland of Sudan and takes the name of Bahr-el-Azraq (the Blue Nile). Two major tributaries, the Dinder and the Rahud, join the Bahir-el-Azraq between Roseires and Khartoum. It then flows another 735 km north before merging with the White Nile at Khartoum. From Khartoum to the Mediterranean Sea, a distance of over 3,000 kilometres, the Nile receives no other perennial tributaries. The Atbara, which joins the Nile 300 kilometres north of Khartoum, only flows when heavy rainfall associated with the summer monsoon occurs in northern Ethiopia although it contributes a significant amount of the total discharge of the river. Despite large seasonal oscillation of its flows, the Blue Nile supplies 86 per cent of the main Nile water.[20]

Within Ethiopia, the Blue Nile basin is the largest catchment, occupying a land area of 366,000 square kilometres – almost a third of the country's land area.[21] The basin covers 45 per cent of the Amhara, 32 per cent of the Oromiya, and 23 per cent of the Benshangul-Gumuz regional states.[22] The annual runoff of 52.6 billion cubic metres represents 50 per cent of the total annual runoff of all rivers in the country.[23] Many of the tributaries of the Blue Nile can be developed to generate electric power and develop irrigation schemes. However, little of the estimated 711,000 hectares of irrigable land in the basin is under cultivation.[24] According to FAO, only 47,000 hectares of land was under irrigation in the Abbay Basin in 2001.[25] Of the potential dozen or more hydroelectric plants that could be constructed in the basin, Ethiopia has only two hydroelectric plants, the Fincha'a and the recently completed Beles.[26] In all, no more than 5 per cent of the Blue Nile water is retained within Ethiopia's boundaries.[27]

Historical precedence has complicated Ethiopia's access to its own Blue Nile waters. For centuries, Egypt and Sudan had appropriated the bulk of the Nile waters, to which Ethiopia contributes about 86 per cent of the total discharge. The 1959 colonial-era agreement signed between Egypt and Sudan gave most of the Nile waters to Egypt, a much smaller amount to Sudan, and nothing at all to upstream countries, including Ethiopia.[28] In the past, both Egypt and Sudan were reluctant to come to terms with the principle of 'equitable and reasonable use' of water by all riparian states, the fundamental international principle laid down in the Helsinki Rules on Trans-boundary Waters.[29] Egypt has consistently based its entitlement to the Nile water on an international water law principle known as the law of prior appropriation. Egypt argues that this law affords it 'historical rights' or 'acquired rights' to the full use of the Nile waters and to oppose upper riparian states that wish to carry out water projects on the Nile and its tributaries. In the past, Egypt has used diplomatic pressure and the threat of force to gain more control over the waters of the Nile and to undermine Ethiopia's interests on the river. But in recent years, upper riparian states are challenging Egypt's continued dominance over the Nile River. Upper riparian countries, including Ethiopia, have embarked on water development projects on the Nile waters that originate in their sovereign territory, arguing that they could use these waters responsibly without causing appreciable harm to the lower riparian states. As shall be discussed later in this chapter, Ethiopia, in particular, seems unwilling to wait much longer. It has already built two dams on the Beles River (a tributary of the Blue Nile) and on the Tekeze River (a tributary of the Atbara River) to generate electric power. It has embarked on building one of the largest dams in the world on the Blue Nile. Ethiopia's determination to use its waters has now brought Egypt and Sudan, albeit reluctantly, to the negotiating table (more on this later in this chapter).

The Tekeze River has many names. Where it starts near Gonder, it is called Guang. The Kunama, who live in the Ethiopia-Eritrea border region, call it Tika. Near the frontier with Sudan, it takes the name of Setit. After about 800 km, it meets up with the Atbara at Tomat in Sudan. The Tekeze River provides nearly 90 per cent of the discharge of Atbara.[30] The river begins near Lake Ashange in the eastern escarpment of the central highlands. It flows west through a steep gorge, circling the Lasta range, and then north and west again around the Semien range. As it runs through the mountains, it receives water from many tributaries. As it reaches its lower course, however, it becomes diminished during the dry season.[31] The Tekeze basin covers an area of 69,000 square kilometres with an annual discharge volume of 7.6 billion cubic metres. The estimate for irrigable land in the basin is a quarter of a million hectares.[32] With the help of the China National Water Resources and Hydropower Engineering, a dam was completed on the river in 2009, which is now generating 300 megawatts of electricity.

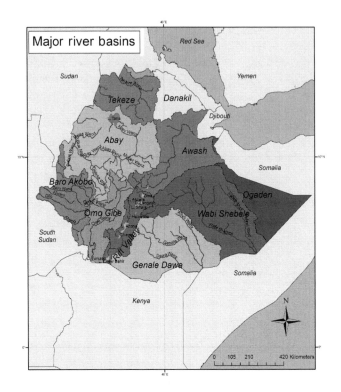

Figure 5.1 Major river basins of Ethiopia.

Map produced by the GeoProcessing Lab at Mount Holyoke College.

The Awash River drains the central and northeastern regions of Ethiopia. This major river originates on the southern slopes of the Worke Mountain Range west of Addis Ababa at approximately 2,500 metres above sea level. Its drainage basin stretches across the northern part of the Ethiopian Rift Valley. The river flows east across the densely populated and intensively farmed Becho plains, where some tributaries join it to create annual flooding during the summer rainy season. The Koka Dam, constructed in 1960, impounds the waters in the Koka reservoir before releasing the river through a dam into the Rift Valley. The steep drop of the river has enabled the construction of hydroelectric facilities at Koka and a series of run-of-river schemes (Awash II and Awash III). The river turns north gradually and flows at some gradient along the foothills of the western highlands. A number of tributaries, including the deep gorge rivers of Kesem and Qabana, join it along this stretch. The Kesem and Qabana rivers cause frequent flooding in middle Awash, as do the Mille and Logiya rivers in lower Awash. Between Koka and the Awash Station, smaller tributaries that drain from the highland catchment to the east also join the river. The eastern plains constitute half of the river's basin area. However, many of the drainage channels in this area never reach the Awash.

The capacity of the river channel is drastically reduced as it passes the Gedebassa swamp located downstream of the Awash Station. The main channel flows adjacent to the swamp and often floods it. Approximately two-thirds of the river's volume enters the swamp. Once the river flows past the swamp, it receives input from the short and very seasonal rivers, the Borkena, Logiya, and Mille, which flow down the eastern slopes of the central highlands. Then, the Awash turns abruptly east and finally terminates in a series of saline lakes, including Gamari, Afambo, Brio, and Abe, near Djibouti about 150 km from the Red Sea coast. These lakes lie below sea level and are often saline.[31] The Awash River flows 1,200 kilometres from its origin to its termination. It is the only major river in the country that does not cross an international boundary.

The total catchment area of the Awash River is about 113,000 square kilometres. Half of this area represents arid lands with no surface discharge. The annual volume of the river amounts to 4.9 billion cubic metres, of which 3,650 million cubic metres are usable.[33] The Awash River valley is the most intensively utilised for irrigation and hydropower generation in the country. The Awash and its left bank tributaries deposit huge quantities of alluvial soils in the river's middle valley, making the area among the most fertile places in the country. The river is also one of the most contested water resources. Government-sponsored irrigation development schemes in the valley have displaced many native pastoral and agropastoral groups (as discussed in Chapter 3). Recent years have seen increased conflicts between groups in the valley stemming from reduced access to pastureland and water resources and a steady increase in irrigation farms.

The Wabi Shebelle is the longest Ethiopian River. Its catchment area of 205,407 square kilometres also makes it the second largest basin in the country. However, it has a lower runoff (3.2 billion cubic metres) than most major rivers because much of the basin is dry land. The river starts in Arsi, near the western ridge of the Eastern Highlands, follows the length of the Gugu and Chercher mountains, and then turns southeast, receiving water from the eastern slopes of the Hararghe highlands. It flows a distance of 1,340 km inside Ethiopia and a further 660 km into Somalia.[34] The average annual rainfall in the basin is 450 mm, with only the northern and western highland parts of the basin, from which the river and its tributaries originate, receiving precipitation in excess of 1,000 mm. Most of the basin has an average annual rainfall of less than 400 mm and decreases in the eastern direction with increased aridity. There are several streams on the east side of the basin, including the Fafan River, but these rivers reach the main river only during heavy rain, which rarely occurs. The Wabi Shebelle never dries up but often disappears short of joining the Juba River in the coastal marshes of Somalia near Mogadishu as a result of evaporation, losses by seepage, and overbank spillage due to low-channel capacities.[35]

The lower valley of Wabi Shebelle has significant economic importance. Large numbers of Somali pastoralists and agropastoralists cultivate crops and herd cattle along the river. Government-sponsored resettlements are increasingly taking place along the river as well. Usually, the Wabi Shebelle River overflows its banks and farmers benefit by practising flood recession agriculture. The irrigation potential in the river basin is estimated at 204,000 hectares.[36] In more recent years, however, the river has been experiencing frequent destructive flooding. Flooding usually occurs when the rains are particularly heavy in the Arsi-Bale-Hararghe highlands. More recent flooding episodes took place in 1996, 1999, 2003, and 2005. Each time flooding occurs, thousands of people and tens of thousands of herds are swept away in Gode, Qalafo, Musthil, and Ferfer weredas – all in the Somali regional state.

The Genale, Dawa, and Web rivers drain the southwestern escarpment of the eastern Ethiopian highlands. The basin has a catchment area of 171,000 square kilometres. The three rivers start just east of the Abaya and Chamo lakes on the eastern slope of the Arero-Sidama-Bale Mountains, which form a divide separating the lakes and the rivers. Because the basin lies nearly entirely within the rain shadow of the Sidama-Bale Mountains, the average annual precipitation is relatively small at 550 mm. Except a small area in the northern mountains, much of the basin receives no more than 400 mm annually, which makes the basin a semi-arid to arid area.[37] The 640-km long Dawa River forms the border with Kenya. Mountain torrents create the Genale, which eventually forms a deep valley after going over some high falls. The Web, flowing down from the southern slopes of Mount Enkolo, also lies in a very deep valley and in some cases even goes underground. The Web meets the Genale, and the Genale in turn merges with the Dawa, near Dolo on the frontier of Somalia, to form the Juba, which eventually drains into the Indian Ocean.[38] The hydrographic catchment area of the Shebelle-Juba basin is shared by Ethiopia, Somalia, and Kenya and extends over an area of about 810,000 km^2. Nearly half of the basin is within Ethiopia, and 90 per cent of the flow comes from runoff in the Ethiopian southeastern highlands.[39]

The Gibe-Omo river basin drains much of southwestern Ethiopia. This river basin benefits from the high precipitation that occurs during the long rainy season from July to October and runoff from the rains from March to April. The source of the river lies in highlands of East Wollega, near Bako, at an elevation of more than 2,400 metres, where it is the Gibe River. The high waters of Gilgel Gibe join the Gibe River from the west, near the mountain town of Abelti. From there, the Gibe is known as the Omo and flows through a deeply incised valley that is broken by many falls. The Gojeb River, which rises from the high rainfall areas to the west of the town of Jimma, joins the Omo in its middle course. After that, the river makes a sharp turn to the west, then southward to drain the lowland valley of the lower Omo, before discharging into the closed basin of Lake Turkana, completing its 1,015-kilometre journey. Lake Turkana is almost entirely in Kenya.[40]

The Omo River has recently become a source of contention and controversy due to the ongoing large-scale, multipurpose water development projects in the valley. The government has built two dams on the Gilgel Gibe. Gibe I generates about 184 MW on a dam with a capacity of 839 million cubic metres. Gibe II, which uses the water from Gibe I through a 26-km long tunnel, generates 420 MW.[41] The government is also building a third dam, the Gibe III, downstream at the confluence of the Omo and Gibe rivers. Construction is already more than one-third done.

When completed, the dam will be the second largest hydroelectric project in the country. It will rise 240 metres high, making it the tallest dam in Africa, and will hold back nearly 14 billion cubic metres of water in the reservoir 150 kilometres long. The dam is expected to generate about 1,870 MW of electricity, more than twice the country's present generating capacity.[42] A fourth dam, Gibe IV, is planned downstream next to the Omo National Park, Ethiopia's largest national park. The dam would have the energy capacity of the other three dams combined, at nearly 2,000 MW.[43] The government plans to export the excess electricity to neighbouring Djibouti, Kenya, and Sudan via a transmission connection.

Environmentalists are concerned that the Gibe III dam will pose a serious threat to the Omo River and Lake Turkana ecosystems. The construction of the dam is feared to disrupt the natural hydrology of the river basin, resulting in severe habitat loss and sharp declines of many animal and plant species. More than 200,000 indigenous peoples of the lower Omo basin are dependent on flood-retreat cultivation, fishing, and grazing livestock along the Omo River. The Omo River collects fertile topsoil from the western highlands and deposits it on a flat plain, and people living along the river use the land to cultivate crops and graze cattle (as discussed in Chapter 3). The dam would regulate flooding and potentially destroy the traditional way of life that has been practised for centuries.[44] The forests and woodlands along the Omo River are also home for a variety of wildlife, including lion, leopard, elephant, kudu, warthog, buffalo, bushbuck, hippo, baboon, colobus monkey crocodile, and numerous bird species.[45]

The construction of the dam might also threaten Lake Turkana's biodiversity. Lake Turkana is the fourth largest lake by volume in Africa. The lake is 260 km long and about 30 km wide, with an average depth of 31 metres and a maximum depth of 114 metres. It has a surface area of 7,560 square kilometres and a volume of 237 cubic kilometres.[46] It is a closed lake without a surface outlet. The Omo River is Lake Turkana's only perennial tributary, supplying more than 90 per cent of the lake's inflow.[47] On the Kenyan side, the Turkwel River drains into Lake Turkana but has been dammed for hydroelectric power generation at Turkwel Gorge west of the lake. The Turkana basin is hot and arid. The average annual precipitation surrounding the lake is less than 200 mm and is quite erratic and unpredictable.[48] Lake temperatures fall between 24.5 and 39 degrees centigrade. Much water is lost via evaporation. At 2,500 milligram/litre, the salinity of the lake is higher than other large lakes in Africa.[49]

Lake Turkana provides a habitat for about 56 fish species, of which 11 (20 per cent) are endemic,[50] including *Haplochromis macconneli*, *H. rudolfianus*, and *H. turkanae*, *Barbus turkanae*, *Brycinus ferox*, *B. minutus*, *Labeo brunellii*, *Lates longispinis*, and *Neobola stellae*.[51] These and other fish species feed the world's largest populations of Nile crocodile (*Crocodylis niloticus*).[52] During the flooding season, which occurs from June to September, various fish species of the lake, including *Hydrocynus forskalii*, *Alestes baremoze*, *Citharinus citharus*, *Distichodus niloticus*, *Barbus bynni*, *Brycinus nurse*, *Labeo horie*, *Clarias gariepinus*, and *Synodontis schall*, migrate up the Omo River to breed.[53]

Other aquatic animals in the lake include *Hippopotamus amphibius*, *Crocodylus* spp., and *Pelusios broadleyi*.[54]

The livelihoods of over 300,000 Kenyan fishermen and agropastoralists depend on Lake Turkana. The Turkana, who are traditional pastoralists, have turned to fishing because of recurrent droughts that have decimated their herd population. The Turkana's favourite catches from the lake include tilapia, Nile perch, and catfish. Environmentalists fear that the construction of the Gibe III dam might damage the ecology of the lake by drastically reducing flow of the water.[55]

The Baro, Pibor, and Akobo Basin, which is about 74,000 square kilometres, is situated in the farthest southwest of the country and has an annual runoff of nearly 23 billion cubic metres.[56] The three rivers combined supply 89 per cent of the Sobat waters in South Sudan.[57] The basin receives the highest average annual rainfall (1,588 mm) of any river basin in Ethiopia. The basin consists of four major catchments: the Akobo, Alwero, Gilo, and Baro. The largest river in the basin is the Baro. The Baro, which is 280 kilometres long, is navigable from the plains of Gambela to the Sudan border, making it the only navigable river in Ethiopia. The Baro follows the boundary for a while before joining the Pibor, which is formed by the Gilo and Akobo. The headwaters of the Baro River rise in the wettest plateaus, south of Gore town, where rainfall totals between 2,000 mm and 2,800 mm annually and the rainy season extends over nine or ten months. The Akobo starts near Lake Turkana and runs northwest along the southwestern border.[58] The basin is characterised by the distinctive features of extensive lowland plain surrounded to the north and east by high plateaus that receive copious amounts of rainfall. The runoff from the surrounding high plateaus inundates the Gambela lowland plain for six to seven months of the year. The water covers an area of about 350,000 hectares overflow.[59] All the rivers, except for the Akobo, are slow moving waters because of the low gradient of the topography through which they flow, allowing aquatic plants to survive in the rivers. Unfortunately, recent years have seen the growth of the invasive water hyacinth in large sections of the Baro River.[60]

Energy and hydropower

Ethiopia has substantial renewable energy resources in the forms of hydropower, solar, wind, natural gas, geothermal, and biomass. Ethiopia's rugged topography is favourable for the development of hydropower. The rivers that cascade down from its highlands have the potential to produce an estimated 650 terawatt-hours per year. In sub-Saharan Africa, only the Democratic Republic of Congo tops this amount.[61] There is a substantial area of geothermal reserves stretching from the Danakil Depression in the Afar regional state along the Rift Valley to the Kenyan border. Geothermal energy can potentially provide about 7,000 megawatts of power. There is a confirmed natural gas reserve of 108 billion cubic metres, the vast majority of which is found in the Somali regional state. Because Ethiopia receives significant sunshine throughout the year, the potential to produce solar energy is vast. Wind energy potential is also great. The wind and solar energy cost almost nothing although the up-front costs are high. All these resources remain mostly underutilised, and per capita energy consumption rate in the country is among the lowest in the world.

Like many other Sub-Saharan African countries, most of the energy consumed in Ethiopia is in the form of biomass. Rural populations almost exclusively use biomass to meet domestic energy demands. Fuelwood, crop residues, and animal dung are all used as fuel. More than two-thirds of all urban households depend on these energy sources as well. Ethiopia produced and consumed more than 93 million cubic metres of fuelwood in 2005, the highest production of timber for firewood in Africa.[62] Very few modern forms of energy are used in the country, and overall energy consumption is low. The dependency on traditional biomass energy sources puts a strain on natural resources; the country is suffering from severe environmental destruction. The combination of population growth, expansion of agricultural land, and high demand for fuelwood for energy has created a detrimental ecological imbalance, including deforestation and soil erosion (as discussed in detail in Chapter 4).[63]

Ethiopia's per capita energy consumption is among the lowest in the world: about 25 kilowatt-hours (kWh) of electricity, 16 kilogram (kg) of petroleum and 276 kg of oil equivalent of other energy sources, mainly biomass.[64] The bulk of energy consumption, about 82 per cent, is in the household sector. Currently, 95 per cent of national energy consumption is from biomass: wood, dung, crop residues, and human and animal power. The remaining 5 per cent is provided by electricity and petroleum products, of which 90 per cent is from hydropower.[65] Access to electricity remains beyond reach for the vast majority of Ethiopians, with only 15 per cent of the population having access to electricity. In the rural areas, where 80 per cent of the population reside, less than 2 percent of the households have access to electricity. In cities and towns, the figures are considerably better, with 86 per cent of the households being serviced, albeit experiencing frequent power interruptions and rolling blackouts. Even with copious rainfall, silting problems in reservoirs frequently diminishes hydroelectric-generating capacity. Ethiopia has the second largest number of citizens without access to electricity in Africa – after oil-rich Nigeria.[66]

The Ethiopian government estimates the demand for electricity will increase by 25 per cent each year. The growth is fuelled by the country's expected gross domestic product (GDP) growth rate at 6 per cent or more, industrial growth rate at 7 per cent, agricultural growth rate at 5.4 per cent, and growth rate in services at 7.7 per cent, urbanisation growth rate at 4.5 per cent, and population growth rate at 3 per cent annually.[67] According to the government's estimate, Ethiopia's power production capacity from hydro, wind, geothermal and solar may well exceed 60,000 MW, which is almost half the total present installed capacity in Africa.[68] The government sees the development of hydropower as the primary means of increasing the energy supply. Ethiopia has the potential to produce 100 times more hydropower than it now does, which is less than 2,000 MW annually.[69] Realising the potential of this resource, the government has designed a ten-year plan to increase the country's power generation capacity about fifteen times the current capacity in 2020. The government plans to invest $13 billion in ten hydropower plants over the next ten years. A 2,000-MW Gibe III hydro dam construction will be completed in 2065. Sinohydro, the Chinese firm that built the famed Three Gorges Dam, has been contracted to build the 1,600-MW Gibe IV dam and the 254 MW Genale-Dawa III and Chemoga Yeda hydropower projects.[70] A 6,000-MW project on the Blue Nile is also under construction. On the completion of these and other hydroelectric projects, the country could become a major exporter of electricity in the Horn of Africa. Then, electricity will easily replace coffee as the top source of foreign currency. The government has already initiated plans for constructing transmission lines to neighbouring Djibouti, Sudan, and Kenya.[71]

There are no comprehensive agreements between the nation-states that share trans-boundary river basins with Ethiopia. Ten states share the Nile Basin, but it is one of the few major rivers in which there is a great disparity between the riparian states. Some states contribute much to the Nile's flow but use very little of its waters (like Ethiopia), while others contribute nothing but use most of its water (like Egypt). The Helsinki Rules laid down equitable and reasonable use of water between riparian states sharing a trans-boundary body of water. However, among the Nile Basin states, reluctance to recognise this principle has remained a challenge in negotiating water use.[72]

Sharing the waters of the Nile River has been a delicate matter of negotiation and political tension for several decades between Egypt, Sudan, and Ethiopia.[73] There is no international treaty for the sharing of the Blue Nile waters between Ethiopia, Sudan, and Egypt. The 1959 Nile treaty between Egypt and Sudan excludes Ethiopia, the source of the Blue Nile. For years, Ethiopia has been pushing the Nile riparian states for an equitable allocation of the Nile waters, arguing that the countries of the Nile Basin have to meet the challenges of poverty, rapid population growth, environmental degradation, and instability. In response to this challenge, the Nile riparian states launched the Nile Basin Initiative (NBI) in 1999. The overall objective of the NBI has been to promote socio-economic development by utilising the Nile's resources. This goal is to be realised through strategic action programmes, including basin and sub-basin joint investment projects. Such projects include promoting efficient use of water for agriculture, implementing trans-boundary environmental protection and rehabilitation projects, undertaking water resource planning and management actions, promoting power linkages and trade between basin states, building confidence between stakeholders, and capacity building through training.

In spite of these positive developments, negotiations through the NBI have been unable to come up with solutions acceptable to all the riparian countries. To date, there is no binding legal and institutional agreement between the states that share the Nile waters that acknowledges each state's right to use its waters or that such rights 'are limited in any way by the principle of just and equitable water sharing.'[74] It is because of the lack of progress in arriving at basin-wide agreements that upstream states have recently begun launching unilateral projects to develop their trans-boundary water resources. Ethiopia, Burundi, Kenya, Rwanda, Tanzania, and Uganda have already signed a Cooperative Framework Agreement declaring colonial era over the Nile waters invalid.

In 2000, Ethiopia issued a proclamation declaring its entitlement to the use of its waters based on international norms and conventions endorsed by the country (Ethiopian Water Resources Management Proclamation, No. 197/2000, dated 9 March, 2000). The Ethiopian parliament also ratified the Nile Basin Cooperation Framework Agreement in 2013, replacing the 1929 colonial agreement that gave Egypt the lion's share of the Nile waters. And finally, despite stiff opposition and dire warnings from Egypt, Ethiopia pressed ahead with the construction of one of the most complex hydraulic projects in the country's history, the US$5 billion dam – the Grand Ethiopian Renaissance Dam (GERD) – over the Blue Nile River (Figure 5.2 shows the GERD project construction underway).[75] The dam inundates 160,000[76] hectares of land and is predicted to be the largest hydroelectric power plant in the African continent and the seventh largest in the world, adding about 6,000 MW of installing capacity to Ethiopia's energy grid. It has the capacity of holding 63 billion m³ of water, twice the size of Lake Tana, which is the largest natural lake in the country[77] or equivalent to the annual flow of the Nile at the Sudanese-Egyptian border. One beneficial outcome for both Egypt and Sudan is that this mega dam will retain much of the silt washed away from the Ethiopian highlands. That will certainly help increase the useful lifetime of the Roseires, Sennar, and Meroe dams in Sudan and the Aswan High Dam in Egypt. Siltation might shorten the lifespan of the GERD unless the country embarks on a massive re-afforestation of the Blue Nile basin.

While Sudan has sided with Ethiopia on the GERD project from the start, Egypt opposed it despite Ethiopia's assurance that there will be negligible reduction, if at all, in the flow of water downstream during the years the reservoir fills up. However, after years of foot-dragging and hostile approach toward dealing with Ethiopia on the use of the Nile waters, Egypt finally came around and chose cooperation over confrontation by acknowledging Ethiopia's right to build the GERD and signed the agreement on Declaration of Principles (DoPs) on the GERD in March 2015. The DoPs agreement, signed between Egypt, Ethiopia, and Sudan, outlines the principles determining the managerial approach that Egypt, Sudan, and Ethiopia should adopt for the Eastern Nile waters.[78] The principles are common understanding, good faith, development, not causing damage, fair and appropriate use of water, trust building, exchange of information and data, dam security, sovereignty, unity and territorial integrity of the state, and peaceful settlement of disputes. By agreeing to and signing of the DoPs, Egypt has accepted Ethiopia's right to use the waters of the Blue Nile. Egypt's President Al-Sisi paid a visit to Ethiopia immediately following the signing of the DoPs in Khartoum. In his address to the Ethiopian Parliament, he praised the Ethiopian government for its understanding, flexibility and genuine mindfulness of Egypt's water needs and acknowledged Ethiopia's right to build the GERD. Egypt's turnaround is a milestone in the history of cooperation on the Nile waters, and the hope is that the

Figure 5.2 The GERD under construction: 40 per cent completed by March 2015 (photo by author).

cooperation between the three lower riparian countries will reduce tensions and build trust for a future comprehensive agreement between all the riparian countries. Also, basin-wide cooperation is vital to sustainable water resource management in the region.

Water flow in the Nile River is predicted to fall by 60 per cent or more over the next three decades.[79] Given this dire prediction, Egypt, Sudan, and the rest of the Nile riparian countries must improve water management and use. For instance, farmers in both Sudan and Egypt use irrigation water wastefully and excessively to the point of causing salinisation and waterlogging of soils. About 30 per cent of Egypt's irrigated lands suffer from waterlogging and salinisation due to overwatering. Water efficiency is abnormally low in both countries because water continues to be highly subsidised. In fact, many farmers in these countries think that water is an unlimited resource and should be available free of charge.[80] The Ethiopian farmers who live in the same river basin suffer from water shortages and unreliable supplies.

Lake ecosystems

Ethiopia has dozens of large and small lakes as well as several major swamps[81] that are found in various ecological zones and at various altitudes. Most lakes of economic importance are in the Ethiopian Rift Valley, including central Rift Valley lakes of Ziway, Langano, Abijata, Shala, and Hawassa (from north to south). Located further south in the Rift Valley are lakes Abaya, Chamo, Chew Bahir, and Turkana – which is almost entirely in Kenya. Seven smaller lakes are scattered around the town of Bishoftu. A group of saline lakes are found in the most arid section of the Rift Valley near the Djibouti border, including Afambo, Laitaf, Bario, Gamari, and Abe. Further north in the Afar lowland are desert lakes like Asale and Karum. Four major lakes – Tana, Ashange, Hayk, and Wonchi – are found in the central and northern highlands of the country.[82]

Lake Tana

Lake Tana sits at 1,820 metres above sea level. This crater-like freshwater lake is the largest body of inland water and the main source of the Blue Nile. The lake accounts for half of the total inland water in the country. The drainage area of the lake is 16,500 square kilometres, of which 3,600 square kilometres is the lake area. It is about 80 kilometres long, 68 kilometres wide, and has a 250-kilometre perimeter. The lake is very shallow; it is only 14 metres at its deepest and, no more than 4 metres to 7 metres deep elsewhere.[83] Thirty-seven islands are scattered about the surface of the lake, half of which house 'ancient churches, monasteries, and rich fauna and flora.'[84] Bahir Dar, the capital of the Amhara regional state, sits on the lake's southern shore. Numerous rivers and streams discharge 10.3 billion cubic metres of water into the lake annually,[85] but about half of the water is lost to evaporation.[86] The Gilgel (Little) Abbay, Megech, Gumara, and Rib are the major contributors. Up to 90 per cent of the region's total rainfall occurs during the June to September rainy season. As a result, the average annual flow at the outlet of the lake is about 3.5 billion cubic metres during the wet season and about 1 billion cubic metres in the dry season.[87] A large area around the lake is wetland. The flood plains of Fogera and Dembia are the largest wetlands adjoining the lake. Heavy flooding during the rainy season contributes to the wetlands and the sediment load of the lake. Both wetlands are important waterbird habitats, including the 'vulnerable Wattled Crane and the Greater Spotted Eagle, the near threatened Pallid Harrier and the Lesser Flamingo and Roget's Rail.'[88] The wetlands also provide habitat for many migratory birds. However, these wetlands are quickly shrinking as water is drained to make room for crop cultivation and grazing.

The most remarkable feature of Lake Tana basin is the presence of extensive agricultural activities and the virtual absence of forests – except on the islands scattered in the lake. Cultivation has expanded up to 3,000 metres above sea level. Where land is not cultivated at higher elevations, including the Choke Mountains, the major natural habitats are moist moorland with Jibrra (giant *Lobelia* spp.), lady's mantle (*Alchemilla* spp.), sedges and tussocks of guasa (*Festuca* spp.), as well as other grasses, montane grasslands, cliffs, and rocky areas. Woody plants, such as asta (Erica), Amija (*Hypericum reolutum*) and bamboo or qerkeha (*Arundinaria alpine*), are only found in patches. Extensive beds of papyrus (*Echinochola* sp.) grow on the shores of the lake. Papyrus is a versatile plant that is used for making reed boats or *tankwas*, roofing, matting, and fencing. There is some vegetation in proximity to the lake dominated by sedges (qetema), reed grasses (shembeqo), and bulrushes (filla) along with swamp grasses like *Echinochla* spp. (asendabo) and *Cynodon aethiopicus* (serdo), which make for very suitable grazing during the dry season. Broad-leaved trees, including Albizia spp. (sas), *Croton macrostachyrus* (bisana), *Cordia africana* (wanza), *Olea europaea* subsp. (surround the churches and monasteries on the islands), *Cuspidata* (weira), *Phoenix reclinata* (zambaba), and figs.[89]

An extensive flat land that is settled surrounds Lake Tana. In the last few decades, population and urban growth have led to increased demand for firewood, expansion of farmland, and industrial and commercial development in Bahir Dar and its hinterlands surrounding the lake. Deforestation in the upper catchment areas and extensive farming activities have caused increased soil erosion, which, in turn, has led to increased sediment influx into the lake. The lake annually receives an estimated 9.6 million tons of sediment load from the upper catchment areas and the many rivers that feed the lake. The outflow of sediment overload from the lake is about one million tons, which means 8.6 million tons of sediment is retained in the lake and the wetlands around it annually.[90]

Water hyacinth – an invasive free-floating aquatic weed – covers large sections of Lake Tana's northern shores in Gonder Zuria and Dembia weredas.[91] Water hyacinth is a fast-growing freshwater plant that can multiply quickly by vegetative reproduction. The plant has broad, thick, and glossy leaves and can rise a metre or so in height. The leaves are 10 centimetres (cm) to 20 cm across, supported above water surface by long spongy stalks. The free-floating mats of hyacinth can cause several problems: it can create fishery losses as a result of competition for light and nutrients, interfere with fishing and navigation, displace native flora, and possibly increase the loss of water through evapotranspiration from the leaves. The weed also poses a significant health threat by providing favourable habitat for malaria-carrying mosquitoes and disease-carrying snails.[92]

The lake also is under threat from pollution and increasing demand for water, owing mostly to growing economic activities in Bahir Dar, one of the fastest growing cities in the country. Commercial establishments, such as hotels, banks, retail shops, small-scale industrial plants, and infrastructural developments, have been on the rise in the last two decades. Many of the establishments built on the lake's shore have found it convenient to discharge untreated waste directly into the lake. Nutrient runoff from nearby farms, discharge of untreated sewage from the surrounding settlements, and manure runoff from grazing lands are slowly enriching the lake. This pollution has had adverse effects on the growth of local fish species, such as tilapia and the livelihood of local fishermen. Pollution has precipitated macrophyte (rooted and floating aquatic plants) on an extensive scale, interfering with the reproduction of the tilapia fish. This dense distribution of macrophyte is due to sediment accretion and the nutrient content of sediments. Excess sediment loading is known to make it difficult for fish species like tilapia to breed. Shallow lakes like Lake Tana are more vulnerable to macrophyte infestation because they are of medium alkalinity and high pH. There is a concern that excess plant growth and increases in phosphorus levels could lead to eutrophication of the lake.[93]

Current conservation policies are inadequate. The lake ecology and the surrounding communities, which depend on it, are under severe threat. The lake provides livelihoods to as many as three million people living around it. Water resources for agriculture and livestock and a significant fishing industry are made possible by Lake Tana. People cultivate large quantities of staple crops, such as rice, pulses, and tef, in the watershed. The Ethiopian Department of Fisheries and Aquaculture reports that 1,454 tons of fish are caught each year at Bahir Dar. It is only 15 per cent of the sustainable catch, according to the Fisheries Department's estimates. The fish resource potential of the lake is about 10,000 metric tons per year. The lake has 26 species of fish, of which 18 (69 per cent) are endemic.[94]

Lake Koka

Lake Koka is an artificial lake located about 150 km south of Addis Ababa. It was the oldest major reservoir built on the Awash River, in 1960, to generate hydro-power, provide irrigation water for the Wonji sugarcane plantation, and control the perennial flooding. The lake covers a surface area of about 180 square kilometres with a storage capacity of 1,850 million cubic metres.[95] Besides the Awash River, the Mojo River is a major contributor of water to the reservoir. The area around the reservoir is densely populated and intensively cultivated for cereals, particularly tef. Farmers also use the alluvial soil around the lake to grow horticultural crops. The wetland around the shore hosts large numbers of water birds, including globally threatened species, such as the Pallid Harrier, Lesser Kestrel, Lesser Flamingo, and Basra Reed Warbler. In addition, the Common Crane, Avocet, Greater Flamingo, migrating Caspian Plover, Yellow-legged Wagtails, and Swallow find a habitat in marshes around the lake.[96] The main fish species in the reservoir is tilapia, but barbs and catfish are also present in smaller quantities.

The lake was once a source of clean freshwater for the thousands of people who live around it. Today it is highly polluted and covered with toxic algae. The source of the pollution is the Akaki River, a tributary of the Awash River. The Akaki River has two main branches. The Little Akaki rises in the northwest part of Addis Ababa, in the eastern slope of Wechacha Mountain. The Great Akaki, the eastern branch of the river, rises from the eastern edge of the Entoto mountain range and drains the northeastern part of the city. Both rivers enter the city as clean, freshwater resources but turn into a toxic sludge after they leave the city and drain into the shallow and hyacinth-infested Aba-Samuel reservoir, about 45 kilometres south of Addis Ababa. The Little Akaki River is by far the most polluted river because several large industries line its banks and tributaries. Nearly nine out of ten of these industries discharge their effluents directly into the river without treatment.[97] The river also serves as a natural sewerage line for domestic wastewater (estimated at approximately 100,000 cubic metres per day) due to the near absence of a sewerage system in the city.[98]

In the last two decades, a number of polluting factories and flower farms have been established along the Awash River and its tributaries, including the Akaki River. These industries also release extremely polluted effluents into the waters that drain into Lake Koka. The Ethiopian Tannery Share Company, in particular, discharges highly polluted effluents into the lake. Agricultural runoffs, human sewage, and factory effluents entering the lake are now causing the growth of notorious, highly toxic algae in the lake. Nitrates and phosphates entering the lake have been especially responsible for the blooming of the toxic blue-green algae known as microcystis. Microcystis cells break open when they die and release the toxin microcystin into the water.[99] Microcystin is a powerful liver toxin; thousands of people who rely on Lake Koka as a source of water have reported severe illness and death from drinking the water.[100] Heavy metal contaminants that are carcinogenic, such as mercury, lead, chromium, and cadmium, have also been found in the water and soils around the lake, indicating risks to human and animal health.[101] Traces of these metals have been found in vegetables grown using the water for irrigation.[102]

There are also other problems for those living close to the lake, such as disease. The marshes around the lake and nearby irrigation schemes have also created a breeding environment for malaria vector mosquitoes. People living in proximity to the lake are 20 times more likely to contract malaria than those living 5 kilometres or more from the lake.[103]

Sediments increasingly fill Lake Koka. Since its impoundment, sedimentation has been an ongoing problem in the reservoir. In the Awash Basin and particularly the Koka watershed, erosion rates are high, often above 6,000 tons per square kilometre per year.[104] An estimated 25 million cubic metres of sediments settle in the lake every year. The lake has now lost 40 per cent of its original water storage capacity of 1,650 million cubic metres because of siltation.[105] This has reduced the reservoir's ability to regulate the flow and to supply water. It is negatively impacting both irrigation projects and any hydroelectric plants that rely on the reservoir for water; nearly 70,000 hectares of irrigated land in the Awash Basin depends on water from the reservoir. Sedimentation can also reduce the reservoir's ability to mitigate floods. Aquatic life in the lake, including fish, is also diminishing fast. As the human impacts of the lake increase through pollution of the lake and inflowing rivers, the capacity of this reservoir to maintain its biological functions will be threatened. There is, therefore, an urgent need to prevent nutrients from runoff of agricultural fields and effluent discharges from industries entering the reservoir.

The Rift Valley lakes

The Rift Valley lakes consist of a series of eight terminal lakes that contain an estimated 5.7 billion cubic metres of water per year. The lakes include, from north to south: the Ziway, Abijata, Langano, Shala, Hawassa, Abaya, Chamo, and Chew Bahir.[106]

Lake Ziway

The open and shallow Lake Ziway sits at an elevation of 1,636 m above sea level. It is the highest, largest, and northernmost of the Rift Valley lakes. The surface area of the lake is about 434 square kilometres, with a maximum depth not exceeding 9 metres.[107] Its water is muddy reddish brown. The small town of Ziway sits on the lake's western shore. Two rivers drain into the lake. The Meki drains the western wall of the Rift Valley, and the Ketar River originates from the eastern side of the Rift and the Arsi highlands. The two rivers combined annually discharge an estimated 675–695 million cubic metres of water into the lake. The lake, in turn, drains southward into Lake Abijata via the Bulbulla River. Approximately 10 per cent of the inflowing water, 110 million cubic metres, discharges into Lake Abijata via the Bulbulla River.[108]

Extensive swamps of *Phragmites*, *Typha*, and *Cyperus papyrus* fringe the northern end of the lake while swamps of *Aeschymene elaphroxylon*, *Panicum repens*, *Nymphaea*, *Potamogeton*, and *Juncus* dominate the western shore and extend far from the shore. There are nine islands scattered in the lake, but many are mostly bare rocks. Three small islands near the southwestern shore of the lake support a thick cover of *Ficus sycomorus*, *Euphorbia*, *Aeschyomene*, several *Acacia* species, and *Typha* swamp on their margins. The largest island – Tulugudu – is covered with scattered *Acacia* and *Euphorbia*.[109]

The Lake Ziway catchment area has seen large-scale irrigation developments since the 1970s and the 1990s. The 2000s have especially seen the development of 7,500 to 10,000 hectares of newly irrigated areas devoted to the production of vegetables, fruits, and flowers, mainly for export. Irrigation has been made possible by extracting water directly from the lake and the lake's two feeder rivers, the Meki and Ketar. Irrigation is estimated to use annually around 150–200 million cubic metres. Water use for domestic purposes has also seen a substantial rise due to increased urbanisation and a growing population. All of these activities have contributed to the reduction of the lake's water level and river discharges.[110] The drop of the lake water level has reduced the discharge of the Bulbulla River into Lake Abijata. In addition, irrigated fields consume about 59 million cubic metres of water annually from the Bulbulla River. During the dry season, the river often dries up before reaching Lake Abijata, which has significantly reduced the level of Lake Abijata that derives about 42 per cent of its water from the Bulbulla River. The Bulbulla River is also a critical source of water in the area and is the only potable freshwater for its 30-km stretch along the semi-arid floor of the rift. The diversion has thus created shortages of water, especially during the dry season.[111]

The reduction of the water level in Lake Ziway could have severe consequences for the fragile ecosystem. Lake Ziway has broad shallow margins that are often fringed with swamp, along with dense floating vegetation and a high concentration of phytoplankton. This environment provides an important habitat for fauna. Lake Ziway has the heaviest fish stock in the region and is the principal source of commercial fishing in Ethiopia. Three endemic fish – *Barbus aethiopicus*, *B. microterolepis*, and *Garra makiensis* – live in this lake, as do the introduced species of *Tilapia zilli*, *Clarias gariepinus*, and crabs.[112] As such, the main economic concern of using water from Lake Ziway for irrigation is how this will affect the fisheries. Another consequence of using water from the lake for irrigation is the effect on the vegetation around the lake's edge. This vegetation provides food and shelter for many animals. Some species are sensitive to short-term disruptions in their environment. Large numbers of bird species inhabit the lake's edges, including cormorants, spur-winged goose, fish eagles, Ibis, egrets, herons, stilts, darts, snipes, grebes, gulls, and ducks, which make the area scenic. Some hippopotamuses inhabit the lake.[113] Irrigation and deforestation have already affected the large mammalian population.

Around Lake Ziway is a rim of grassland that provides an important source of dry-season grazing for livestock. If the level of the lake drops, there may be an increase in transpiration loss from the marginal vegetation. This in turn would cause the groundwater table to drop, endangering the grassland. Lowering the groundwater level would also endanger the water supply of the local community, as the springs on the eastern shore of Lake Ziway would dry up.[114]

Lake Ziway is receiving increased sediments and nutrients via its feeder rivers that originate in the intensively cultivated highland areas. Fertiliser use has been growing in recent years. It will inevitably increase the amount of nutrients entering the lake, thus resulting in eutrophication. Eutrophication promotes the growth of algae and other aquatic plants. One of the main causes of algae growth is an excess of the nutrients nitrogen and phosphorus. These nutrients not only come from agricultural activities but also from industrial and municipal sources. Significant quantities of these nutrients could cause the phytoplankton population to increase with numerous undesirable results. The colour and odour of the water would change. Phytoplankton and other aquatic plants would settle at the bottom of the lake and create a sediment oxygen demand. The oxygen dissolved in the water would then drop. The growth of rooted macrophytes, or larger plant forms, would interfere with navigation and aeration. Human activities in the catchments around these lakes may ultimately result in fish kills, algal blooms, and the death of wildlife.[115]

If the water level of Lake Ziway falls enough to make it a terminal lake, its salinity levels will rise enough that, within a short period, lake water will be too saline to drink or use for irrigation. As a shallow lake, averaging only 4 metres in depth, it is very susceptible to water loss from irrigation. As the largest water resource in the Central Rift Valley, its loss would be significantly damaging to Lakes Abijata and Shala in particular and the Abijata-Shala Lakes National Park in general. The diversity of birdlife within the park is vast, with over 400 species recorded. The park is beneath the major flyway for migratory birds, especially raptors, flamingos, and many other water birds. The edges of Abijata are ideal for wading birds and ducks, as there are fluctuations in alkalinity, creating the algae that attract food for these shorebirds. These alkalinity levels are determined by discharge from the Bulbulla River that is fed by Lake Ziway.[116]

Lakes Abijata and Shala

Situated at an elevation of 1,573 metres, Lake Abijata is a relatively small, shallow, alkaline, closed terminal lake. The lake is 17 kilometres long, 15 kilometres wide, and has a surface area of 205 square kilometres with a maximum depth of 14 metres.[117] A steep escarpment along its eastern shore and gently rising land along its western and northern shores surround the lake. Steeply rising hills separate it from Lake Shala to the south. Pasture, mainly of salt adoptable *Sporobolus spicatus* and *Cynodon dactylon*, dominates its northern shore, while Acacia varieties, *A. Seyal, A. campyacantha, and A. Tortilis*, surround the lake elsewhere. Patches of *Cyprus, Ficus sycamores, Ricinus communis, Acacia albida, and Crotalaria laburnifolia* occur along the Bulbulla River. There is, however, clear evidence of overgrazing, soil disturbance, and deforestation surrounding the lake.[118]

Three rivers feed into Lake Abijata: Gogesa, Hora Kelo, which originates in nearby Lake Langano, and the Bulbulla River, which drains from Lake Ziway. Over time, the level of Lake Abijata has dropped drastically and is shrinking (as can be seen in Figure 5.3). Lake Abijata's shallow depth and terminal position make it more susceptible to climatic changes, variations in precipitation, and the flow of its feeder rivers. As it is a closed lake, its only significant water loss is through evaporation. Groundwater flow from the lake is negligible. In general, changes in the water level are related to variations in rainfall and river flow. However, recent schemes to pump water for soda ash production and the withdrawal of water from the rivers that feed it and upstream Lake Ziway have begun to affect the water level. When irrigation happens year-round, its effect on the lake is magnified as water continues to be extracted during the dry season when evaporation rates are high and precipitation rates are low.[119] At present, vast salt plains border the lake. Increasing silt loads are also endangering the lake because the rivers feeding into the lake are being heavily used for agricultural purposes.

The establishment of a soda ash plant on the shore of Lake Abijata is also not without adverse consequence. The plant produces 'soda ash (Na_2CO_3) from sodium bicarbonate ($NaHCO_3$) dissolved in lake water.'[120] The plant evaporates lake water and then collects the remaining sodium bicarbonate. It is then heated and decomposed into water, carbon dioxide, and sodium carbonate (soda ash). The plant annually produces 20,000 tons of sodium carbonate.[121] Water extraction from Lake Abijata for the production of soda ash is about 13 million cubic metres each year.[122] The extraction threatens the biodiversity and increases the salinity and alkalinity of the lake.

Lake Abijata is of great biological importance for a number of reasons. The lake is shallow and alkaline and has a muddy shore that supports a wealth of bird life almost unparalleled anywhere else in Africa. The Ethiopian Rift lakes also form an important migration route for Palearctic birds during the northern winter. Lake Abijata is also part of the Rift Valley Lakes National Park. The blue-green algae of the lake support a large flamingo population, while many other birds rely on the lake for fish. Lake Abijata is a vital feeding ground for Cape Wigeon, Abdim's Stork, and Great White Pelicans. Great White Pelicans breed in Lake Shala, in large numbers. However, the alkalinity in Lake Shala is too high to support a large fish population for the pelicans to feed on. The pelicans, therefore, rely on Lake Abijata as their feeding grounds. The increased changes in the alkalinity of the lake have caused a reduction in the fish population, ultimately leading to the death of fish-eating birds.[123]

At an elevation of 1,558 metres, a mere 2 km south of Abijata is Lake Shala. The lake is 28 km long and 12 km wide, with a surface area of 409 square kilometres. Its maximum depth of 266 metres makes it the deepest lake in the country.[124] Its water is blue, clear, and quite stormy in contrast to the other lakes. Hot springs (90–100 degrees centigrade) gush from the banks and bottoms of the lake.[125] Two perennial streams and many seasonal creeks drain into the closed basin of Lake Shala. Open Acacia vegetation and a belt of *Euphorbia abyssinica*, *Typha*, and *Sesbania punctate* grow to the east of the lake. Patches of similar vegetation also cover the two largest islands in the lake (see Figure 5.4).[126]

Figure 5.3 Lake Abijata is shrinking in size and depth. The foreground used to be part of the lake.

Figure 5.4 Dwindling acacia forest fringing Lake Shala (photo by author).

Lake Shala's shores are steep, and its waters are deeper and more alkaline than Lake Abijata. As a result, the lake produces a limited amount of biomass to support aquatic lives. There are still many bird species, however, and four of the lake's nine islands are used as breeding sites by many birds, and are home to the continent's most important breeding colonies of Great White Pelicans, Cormorants, and Storks. A tiny saline crater lake, Chittu Hora, sits less than 2 kilometres east of Lake Shala and is also a home to 5,000–10,000 flamingos.[127] Much of the acacia woodland surrounding the lake has been cut down and turned into charcoal and cattle are grazed under scattered acacia trees.

Lake Langano

To the east of Lake Abijata, separated by the Addis Ababa-Hawassa motor road, is the resort Lake Langano. The lake is 18 kilometres long and 16 km wide, with a surface area of 230 square kilometres and a maximum depth of 46 metres.[128] Like Ziway, its water is muddy and reddish brown. Its shores are rocky in the east, swampy in the north, and sandy in the west. A steep escarpment of 20 metres or more borders most of its western shore. The lake is drained by the Hora Kelo River, which empties into the adjacent Lake Abijata. The lake receives the bulk of its waters from several small streams – Jirma, Gedemso, Lepis, and Huluka – that originate in the Arsi highlands to the east and south. A fifth of its annual water also comes from precipitation. The water of Lake Langano is more stable than Ziway or Abijata. There is no major irrigation within the Langano catchment although pastoral and agropastoral activities are increasing around it. The lake is kept stable by groundwater flow coming from springs and large underground faults.[129] Hot springs awash its northern shore.

Dense acacia woodland (mostly *A. seyal*, *A. tortilies*, and *A. campylacantha*) and bushes covered the areas to the north and west of the lake only three decades ago. But today much of the vegetation has been cleared, and a large number of pastoral and agro-pastoralists have settled in the areas. *Sedges*, *Ficus sycomorus*, *F. vasta*, and acacia vegetation cover much of the rocky, eastern side of the lake. The northern and southern margins of the lake sustain *Typha*, *Juncus*, and *Sesbania* marshes. Farther up the slope of the Arsi Mountains in the southeast is the Munissa State Forest, with *Typha*, *Sesbania*, *Ficus*, *Maytenus*, *Croton*, *Acacia*, *Combretum*, *podocarpus falcatus*, *Hagenia*, *Hypericum*, *Erica*, and *Alchemilla* being the most dominant vegetation, in ascending order.[130] Colobus, Olive Baboons, and Vervet Monkeys inhabit this forest. The lake provides a home to large numbers of water bird species.[131] Along with Lakes Abijata and Shala, Lake Langano also provides excellent ground for migratory birds from Europe and Asia, as well as breeding and roosting sites for Flamingos and Pelicans.[132] Tilapia and barbus are significant fish stocks in the lake.

Lake Hawassa

Moving south for about 100 kilometres from Lake Langano is the smallest of the Rift Valley lakes, Lake Hawassa. With a surface area of 129 square kilometres and a drainage area of 1,259 square kilometres,[133] it is 16 km long and nine 9 km wide. It has a maximum depth of 10.7 metres.[134] The immediate surroundings of the lake are flat, but the small Tabor Hill rises in proximity to the lake in the northeast, and there are large ranges some kilometres to the northwest. An extensive marsh of *Juncus* and woodland dominated by *Maytenus senegalensis, Rhus natalensis,* and *Balanites aegyptica* borders the lake's shores.[135] The city of Hawassa, with a population of 157,000, occupies the eastern shore of the lake. A 2-kilometre dike runs along the eastern shore of the lake to prevent flooding the city when the lake swells up, but the structure appears to be too short to contain the water. Drawing excess water from the lake for irrigation or channelling it into nearby drainage systems is other ideas being considered to prevent potential overflow.[136]

The lake does not have an outflow, and probably has a subterranean inflow, judging from its relatively diluted nature, as well as dilution from the feeder Tiqur Wuha River and basin overflow. The lake water is home to over 100 species of phytoplankton, freshwater invertebrates, bacterioplankton, and six different fish species, three of which are both endemic and commercially important. The lake is also the habitat for some species of waterfowl and reptiles. The Shallo Swamp drains into the lake. The lake is extremely biodiverse due to its plants, animals, and microorganisms. Emergent and submerged macrophytes cover the littoral area, serving as shelter and breeding zones for weed bed fauna (annelids, crustaceans, and insect), protozoan, rotifers, cladocerans, copepods, ostracods, and fish. Some of these plants include *Cyperaceae, Typha, Paspaladium, germinatum, Nymphaea coerulea, Pistia stratiotes,* and *Wolfia arrhiza.* The lake also supports over a hundred species of both local and migrant Palearctic water birds.[137] Hippopotamuses, otters, monitor lizards, and vervet, grivet, and Colobus monkeys are also in the habitat.

The ecosystem in its entirety stabilises the microclimate and also provides cultural and religious services to the local people. Lake water uses include irrigation, bathing, recreation, and drinking water for both humans and animals. The lake supplies the city of Hawassa and the surrounding communities with all their water and supports a flourishing local fishery. The fish (mostly *Barbus, Clarias,* and *Tilapia*) are used both for home consumption and for the market, while the vegetation is used for grazing, boat making, mattresses, mats, agricultural implements, and building construction.[138]

Human activities have caused a range of serious ecological problems for the lake, river, and swamp. Urbanisation, irrigation, forest clearance, factory construction, and the use of fertilisers, herbicides, and pesticides have contributed to the damage of these ecosystems. Leached chemicals from both large- and small-scale farming are reaching the lake by drainage systems. As the number of hotels, commercial establishments, and factories grow, the amounts of solid and liquid wastes being generated are also growing. Most of the town's sewer lines terminate in the lake, and the municipality does not currently have a system to manage or collect most of its solid and liquid wastes. Another major problem is the untreated toxic discharge that comes from many industries in the area, including state-owned industries of textile, flour, ceramic, sisal, and tobacco production. The Hawassa Textile factory uses large quantities of chemicals, water, and energy and discharges

Figure 5.5 Lake Hawassa: eutrophication in process (photo by author).

50 cubic metres of wastewater per hour at less than full capacity. This waste is coloured and high in biochemical oxygen demand (BOD), chemical oxygen demand (COD), and total suspended soils. It is highly alkaline, with a pH of 12, and hot. The factory releases untreated waste directly into the Shallo Swamp, the effects of which transfer into the hydrological system of Lake Hawassa and affect both the ecosystem and the societies that rely on it for food, water, and livelihoods. Solids from the fibrous substrate and processed chemicals can hinder oxygen transfer and reduce light penetration. If they settle on the wetland beds, they can cause the formation of an anaerobic layer. The BOD varies considerably in type. Some, such as starches, are entirely biodegradable while others, such as the refractory compounds found in dyes, are entirely non-biodegradable. Effluents with high concentrations of xenobiotic deplete the dissolved oxygen (DO) in water, which kills fish and reduces water quality. Phosphates entering the water from detergents can cause the nutrient levels in the water to rise and leads to algal blooms, which, in turn, can alter the food chain and ionic composition of the water, increase organic matter in the sediment, decrease metalimnetic and hypolimnetic oxygen (which causes fish suffocation), and cause changes in the water temperature. The water of the swamp is causing illnesses and is no longer considered potable.[139] Eutrophication of the lake is in process (as Figure 5.5 shows).

Lakes Abaya, Chamo, and Chew Bahir

Lakes Abaya and Chamo – which sit side by side – are located farther south in the Rift Valley. Savannah plains and mountains surround the lakes. The catchment of the lakes' basin covers an area of about 18,000 square kilometres. The town of Arba Minch sits in the foothills of the western Rift Valley escarpment, overlooking Lake Abaya to the north and Lake Chamo to the south. At the base of the escarpment, below the town, is a flat land covered with dense evergreen forest that provides a habitat for many mammals and birds. Numerous small springs bubble up from the forest floor, perhaps the 'only groundwater forest in East Africa.'[140]

With a water surface area of 1,160 square kilometres, Abaya is the longest and second largest lake in Ethiopia. The lake is 60 km long, 20 km wide, and has a maximum depth of only 13 m. Low vegetation of grasses and sedges cover the shores of the lake, and away from the lake shore acacia-commiphora woodland is the dominant vegetation.[141] The lake is fed by numerous mountain rivers, of which the Gidabo and Billate flow year-round.[142] The Billate River rises from the southern slope of Mount Gurage, while the Gidabo drains the eastern slope of the Guge Mountains. Lake Abaya is reddish in colour, owing to a vast amount of suspended sediments entering the lake, mainly via the Billate River.[143]

The blue water of Lake Chamo is 26 kilometres long, 22 kilometres wide, and has a maximum depth of 10 metres.[144] The lake receives regular inflows of water via the Kulludu River and several small streams. In addition, Lake Abaya overflows into Lake Chamo via the Ualo River, especially during years of unusually heavy precipitation in the highlands surrounding the lakes. However, Chamo itself overflows into the Sagan in years of heavy rainfall. The 551-square kilometre Lake Chamo is separated from Lake Abaya by a kilometre-wide ridge surrounded by swampy forestland in the west. To the east of the ridge is the savannah plain of Nech Sar, a national park established as a sanctuary for the surviving herds of Swayne's hartebeest, Burchell's zebra, and Grant's gazelle. The park also protects large parts of Lakes Chamo and Abaya.

The lakes provide a habitat for many species of fish, including tiger fish, giant Nile perch, barbel, catfish, and tilapia. A large number of hippos and waterbucks make their home in Lake Chamo. The lake is also teamed with Nile crocodiles. The forest and savannah lands that surround the lakes provide home to Colobus and velvet monkeys, olive baboon, warthog, bushbuck, and bush pig. The forest provides habitat for numerous bird species, some of which include Kingfishers, Great White Pelicans, storks, ibises, silvery-cheeked hornbills, cormorants, and black and white fish eagles.[145]

Abaya and Chamo suffer less impact by anthropogenic activities than most other lakes. There exists no large extraction or diversion of water from the lakes. Rivers and streams drain into the lakes unhampered. They do not experience significant seasonal variation in precipitation. Pollution is a negligible problem mostly because of the limited degree of urbanisation and industrialisation in the areas surrounding the lakes. There is, however, a concern that wastes from the Arba Minch textile industry are contaminating Lake Abaya since there are no restrictions on disposing of wastes into watercourses. Local people, including the Ganjulle, Gujji, and others, utilise the resources of the rivers and lakes with little or no adverse impacts on their ecology. However, the last few decades have seen increased sedimentation of Lake Abaya. Sediment inputs to the lake come from inflowing rivers and the rain-induced erosion from farms and areas devoid of vegetation cover.

Chew Bahir is a vast saltwater lake located at the southern end of the Ethiopian Rift Valley. The southern point of the lake lies within the Kenyan territorial space. The lake is 64 kilometres long and 24 kilometres wide. The Weyto and the Sagan rivers are the main bodies that feed into Lake Chew Bahir. The Bala and Dancha tributaries of the Weyto drain the northern and western part of the catchment. The Sagan River rises in the Amaro Mountains, and Lake Chamo drains the eastern catchment. The Weyto and Sagan rivers join to form the Gelana Delai River about 50 kilometres north of Chew Bahir.[146] Some decades ago, the lake covered about 2,000 square kilometrers, with a maximum depth approaching 8 metres. Today, it has significantly shrunk into a 'seasonally flooded swamp and salt marsh.'[147] The lake has no outlet and lies in an area where evaporation (2,000 mm/year) exceeds precipitation (500 mm/year) four-fold.[148] As a consequence, the lake almost entirely dries out for most of the year. The lake swells up occasionally with substantial inflows from the rivers and streams that feed into it. The water of the lake is highly saline for human and animal consumption. The Hamer, Arbore, and Tsemay inhabit the Weyto River valley and Chew Bahir basin. They are agropasto-ralists who keep small animals and cultivate some crops when climatic conditions permit. But they suffer from a constant shortage of and conflict over access to freshwater springs in and along the margins of the lake.[149]

The lake and its hinterland are part of the Chew Bahir Wildlife Reserve. The lake's salt-tolerant vegetation and dry acacia savannah woodland in the catchment support a fair number of Grevy's zebra, greater and lesser kudu, gerenuk, Grant's gazelle, Oryx, spotted hyena, among others. The perennial swamp that lies at the mouth of the Gelana Delai River provides a home for thousands of bird species.[150]

In sum, development activities have an effect on the Ethiopian lake ecosys-tems, such as land and water development, pollution, the introduction of exotic species, and the overexploitation of fish stocks. Some aquatic biota cannot tolerate these conditions, while others thrive on them. Other activities with serious effects include damming and diversion of rivers, channelling and building water distri-bution facilities, and indirect influences, such as the removal of vegetation cover in drainage basins for land. Negative effects, such as changes in hydrological systems, temperature conditions, nutrient levels, and sediment deposits in dams, have all been observed. River and stream diversion reduces habitat diversity, while canals can be vectors for pathogens. Removal of vegetation cover causes increased erosion and sediment in water bodies. Soluble fertilisers can pass into the water from agricultural land, leading to increased nutrient loads and the accumulation of toxic chemicals. Other threats include the introduction of organic sewage and domestic wastes, as well as detergents and other pollutants, possibly causing the demise of freshwater crabs. Perhaps due to irrigation schemes, water levels have also decreased, having severe effects on the fish populations. The introduction of exotic species is also widespread.[151]

Wetlands

A wetland is a geographic area that has the features of both land and a body of water. They usually occur in low-lying areas that receive water from adjacent surface water. This may be freshwater from lakes, rivers, ponds, or streams. It may also be saltwater from tides in coastal areas. In wetlands, the water table is usually at or just below the ground surface for an extended period. This creates an environment in which only plants adapted to wet conditions can thrive. It also promotes soil features like those found in wet environments. The existence of wetlands, as well as their biological components, is determined by water.[152] In Ethiopia, this ecosystem contains marshy areas, swamplands, floodplains of major rivers, natural and artificial ponds, margins of lakes, high mountain lakes, and upland bogs. This ecosystem is often thought to be a wasteland, with agricultural or pastoral value only after draining. The wetlands are often considered sources of disease.[153]

The exact areal coverage of wetlands in Ethiopia is unknown. Few detailed studies on wetland resources exist. Areas covered by wetland change seasonally. They shrink during the dry season and expand during the wet season. Johnston and McCartney estimate that around 1.5 per cent of the total land area of the country may represent wetlands,[154] while others estimate 2 per cent or 22,500 square kilometrers.[155] It may reach 5 per cent in the wetter southwestern parts of the country, such as Ilu Ababor.[156] Wetlands found in Ethiopia include alpine formations, riverine, lacustrine, palustrine, and floodplain. These ecosystems support many species of flora and fauna, though in-depth surveys of these have been somewhat lacking. Because wetlands are found at different elevations and climates, the habitats and species within them fall into a wide range of plants, birds, and other animals, including many endemic species.[157]

Because they combine the characteristics of both terrestrial and aquatic environments, wetlands are some of the most diverse ecosystems. Wetlands are habitats for many plant, invertebrate, fish, and large animal species. Many threatened or endangered species also live in wetlands. Wetlands provide homes for species that can live on dry land, those that live in the water, and those that can only live in a wet environment. Many fish species use wetlands as spawning or feeding grounds or as protection from predators. Birds often also rely on wetlands for food, resting, or as a nesting site. Wetlands provide shelter for breeding or migratory birds. There are 73 important bird areas in Ethiopia, of which 30 (41 per cent) are wetlands. This indicates the importance of wetlands as bird habitats.[158]

Inland wetlands can control floods by storing excess water during wet periods and releasing it after the peak flow has passed. When wetlands are situated over pervious strata, they can serve as sites where surface water can seep into the ground and recharge groundwater and springs. Springs can also emerge from wetlands, providing a source of water for animals and people.

Tourists use wetlands as recreational sites for activities such as bird watching, hunting, and fishing. The Rift Valley lakes, such as Abijata, Shala, Langano, Ziway, Beseka, Hora, Bishoftu, Hawassa, Abaya, and Chamo, provide rich sites for scientific study. In the north, Lakes Tana, Hayk, and Ashange provide important habitats for waterfowl, fish, and crocodiles. Because these wetlands are distributed at different elevations in different climatic zones, they provide homes for a wide array of species.[159]

In southwestern and northwestern Ethiopia, wetlands are important resources for communities. They are cultivated to grow maize and vegetables and, thus, contribute directly to food security. Harvests from these areas are usually early in the rainy season, just as food harvested from upland areas is beginning to run out. Crops from wetlands are an important source of food security, especially for households whose upland harvest was poor. Wetlands also contribute to food security indirectly by providing products that can be sold for cash. Some poor households make a living collecting materials from wetlands to make crafts to sell in the market. People collect medicinal plants from wetlands for domestic use or sale. Wetlands also provide grazing grounds for cattle. Many rural communities also rely on wetlands for drinking water. Wetlands help maintain the water table and local water supplies.[160]

Wetlands cover a large space in Illu Ababor Zone of Oromiya. The zone is in the southwest highlands on a moderately dissected plateau. The rainy season is ten months long and, as a result, it has the highest rainfall in the country. Tropical montane forests, which include wild coffee (*Coffee Arabica*), are predominant. A network of streams and rivers, some of which form wetlands – mostly spring-fed, valley-head swamps, or mid-valley swamps – drains the plateau. About 1.4 per cent of the zone is a wetland, but three of the 13 weredas (Metu, Chora, and Yayu) are more than 2 per cent wetland.[161] Larger floodplains are also common. Vegetation such as the sedge *Cyperus latifolius* or *cheffe* and the swamp palm *Phonex reclinata* are common in these ecosystems. The zone was sparsely settled, and only a third of its surface remained forested until the 1960s. The zone's wetlands played an important role in the livelihoods of rural communities. Their environmental role included hydrological and ecological benefits, such as the recharge and discharge of groundwater, flood control, and sediment retention, and the support of high levels of biodiversity of specific flora and fauna. The socio-economic benefits they provided included the provision of clean water supplies, the production of *cheffe* (that is used for roofing and craft material), and the production of other plants with medicinal benefits. The wetlands serve as moisture reserves during the dry season and have been used on a small scale for early season agriculture.[162]

Demands for wetland use have increased significantly since the 1960s since the Illu Ababor region was slowly brought into the country's economic life through coffee and spice production, in-migration, resettlement, and market-oriented smallholder agricultural development. Forest clearance increased with the growing population. Wetlands were also gradually encroached upon for agriculture as land became scarce and food shortages were more severe. The Derg regime's policy of promoting regional food self-sufficiency meant that it was necessary to develop new ways of cultivating food crops in the Illu Ababor region. The government viewed wetland cultivation as a way to produce food during a season that usually had food shortages, and drainage allowed the area to meet its requirements under the policy. Coffee production was significantly promoted by the Derg government due to its value as a source of foreign exchange. Farmers were encouraged to plant coffee in the forests near their cultivated land, which partially restricted forest clearance even as the population grew, and soil fertility decreased. Farmers then turned to wetland drainage for alternative land. Other Derg policies also put pressure on the wetlands. After the 1984 famine, the government resettled half a million people from the north to the more watered southern areas. The government pursued integrated resettlement in Illu Ababor to link settlers to established communities, which were then expected to provide land and support for the new arrivals. It sometimes allocated wetlands, as they were considered less valuable than forestlands; this increased wetland cultivation significantly in the late 1980s though many settlers returned to the north after the 1991 change of government. Villagisation, the late 1980's process in which the Derg regime attempted to concentrate the rural people into villages, also affected the wetlands. The sudden demand for building materials, particularly *cheffe*, put severe pressure on the wetlands near the new villages and caused the reduction of many sedge beds.[163]

Though most wetland cultivation has remained sustainable, the above pressures have caused a shift to unsustainable practices and the degradation of wetlands. As the need for output rises, most wetlands are no longer being restored to their natural flooding regimes, thus exhausting soil fertility and lowering the water table to levels below those able to sustain agriculture. Though farmers acknowledge the effects of their practices, the need for food security is their priority. As the ecological characteristics of the wetlands changed due to the dropping of the water table, the availability of natural vegetation and clean water has declined. After these changes, the wetlands were used as pasture land, but the benefits were short-lived since the cattle aided in soil erosion. These losses are having severe effects on local communities. Dried up springs force women to walk farther to find water, and wetland plants used for medicinal purposes and craft industries have become scarce.

The natural sedge used for roofing also disappears where people completely drained wetlands. Those who are not wealthy enough to afford other roofing materials must then search farther in wetlands for sedge. The grass could be used as a replacement but cannot withstand heavy rains. Even if other communities allow some reed collection, they will not allow those from other communities to collect the reeds for sale, or for uses beyond roof thatch. This has an impact on those poor people who rely on reed cutting and selling to supplement their incomes to provide enough food. Local craft activities are also affected by reed shortages, and these pursuits are important for poorer women to supplement domestic resources. Other wetland products such as palm fronds or medicinal plants also become scarce, with heavier effects on the lower end of the socio-economic scale.[164]

Less than 10 kilometres southwest of Jimma (Oromiya), along the Jimma-Seka-Shebe road, is the Koffe wetland. It is the home of the endangered Wattled crane. It is surrounded by *Eucalyptus grandis* covered low hills that serve as a watershed for small rivers like Melka Faki and Kilo. Agam (*Carissa edulis*), sembelet (*Hyparrhenia* sp.), and serdo (*Cynodon dactylon*) are the dominant vegetation in the wetland. People have intensively grazed the land covered by serdo. The Ethiopian Wildlife and Natural History Society survey indicates the existence of nearly 80 bird species in the area. However, farm and grazing encroachment, deforestation, and soil mining for brick making are threatening the ecological integrity.[165]

In the north, the wetlands around Lake Tana have a high diversity of water birds, including 'the piscivorous little grebe (*Tachybaptus ruficollis*), great white pelican (*Pelecanus onocrotalus*), great and African cormorants *(Phalacrocorax carbo* and *P. africanus*), and darter (*Anhinga rufa*).' Large numbers of Palearctic migrant water birds also rely on the lake and the swamps around it for feeding and resting grounds.[166] However, much of the wetlands around Lake Tana are losing ground to farming activities and grazing, owing mainly to increased population and pressure on agricultural land. For instance, much of the Fogera floodplain wetlands in the eastern margins of Lake Tana are being used for rice farming.[167]

Further east in the South Wollo Zone of the Amhara regional state, about 35 kilometres south of Kombolcha, is the Borkena wetland covering 12,000 hectares, consisting of '7,000 hectares of permanent swamp and 5,000 hectares of annually flooded land.'[168] The wetland is replenished by the seasonal overflow of the Borkena River and many underground springs perennially. This ecosystem provides a habitat for many bird species, including the saddle-billed stork, marabou, the endemic blue-winged goose, ibises, and pelicans.[169] People use reeds and sedges collected from the swamp for thatching roofs and building fences. The land adjoining the wetland is under extensive use for dry season grazing by lowland agropastoralists after the floodwaters recede.[170] Encroachments by smallholders as well as commercial farmers pose a significant threat to the survival of this small but rich wetland ecosystem.[171]

About 20 kilometres south Addis Ababa is the Akaki wetland, a narrow stretch of land alongside the Akaki River, extending for about 13 kilometres between the town of Akaki Beseka and the artificial Lake Aba Samuel. It is a seasonal wetland that only occurs following the commencement of the summer rainy season and is almost completely dry during the dry season. The dominant vegetation includes reeds, tall sedges, and grasses. The wetland provides habitat for flamingos, waders, ducks, and European migrants, including stork and grey and yellow wagtails. The red-chested wheatear, an endemic terrestrial bird, is common in the area, as are the crane and Wattled ibis.[172] Downstream is Lake Aba Samuel, a body of water that was impounded by the Italians in 1939 to generate the first hydroelectric power in the country, but has now turned into a wetland due to nutrient-rich sediment buildup. The lake is entirely carpeted by the fast-growing water hyacinth (*Eichhornia crassipes*) and is devoid of any aquatic life.

Shallo swamp or wetland is a drainage area for runoff from Lake Hawassa's catchment areas before it flows into the Tikur Wuha. It covers a surface area of 77 square kilometres and is inhabited by diverse species of waterfowl, weed bed fauna, zooplankton, phytoplankton, fungi, bacteria, amphibians, reptiles, and macrophytes. The wetland grasses are used for grazing and other domestic needs. The main fish species in this wetland, *Clarias gariepinus*, serves as a source of food for the people. The wetland maintains the hydrology of the area, stabilises the microclimate, and serves as a buffer against harmful additions from natural processes and human activities in the catchment.[173] However, this wetland is being degraded by harmful human activities, such as land use and modification, industrial discharge, and other activities associated with urbanisation.

The Chomen and Fincha'a swamps or wetlands are on the littoral zone of the Fincha'a reservoir that was built in the East Wollega zone in Oromiya in 1973. The reservoir displaced 3,100 families or 14 per cent of the people of the watershed.[174] The reservoir initially contained about 185 million cubic metres of water, but its capacity was increased to 460 million cubic metres in 1987 by diverting the Amarti River into the lake to increase hydropower production. A number of small streams also drain into the reservoir, but direct precipitation is a major source of its replenishment. The inundation of this 239- square kilometre reservoir did render about 30 percent of the households in the area landless. Additionally, the construction caused the loss of 120 square kilometres of wetland, 100 square kilometres of grazing land, 18 square kilometres of cropland, and about 1.2 square kilometres of forest-land.[175] Much of the elevated areas of the watershed are heavily cultivated and grazed and, as a result, about 63 tons of topsoil per hectare are lost via erosion per year. This, in turn, is causing sediment buildup in the reservoir.[176]

Both floating and standing vegetation cover the Chomen and Fincha'a wetlands. The perennial floating stoloniferous grass (*Panicum hygrocharis*), sedges, blue water lilies (*Nymphaea coerulea*), and knotweeds (*Persicaria* spp.) form the dominant vegetation of the wetlands. The wetlands provide habitat for many bird species, including the globally threatened Wattled Crane, Rouget's Rail, Blue-winged Goose, Abyssinian Black-headed Oriole, Black Eagle, Black-crowned Crane, and Lammergeyer.[177] Increased siltation and discharges of chemical ferti-lisers from the surrounding farms pose a significant threat to the ecology of these wetlands.

There used to be large areas of wetlands along the middle and lower courses of the Awash River that provided habitats, vegetation, and water for humans and wildlife. Extensive development of irrigated large farms in the area has vastly reduced available wetlands and vegetation. People and domestic animals subject the few that do exist to intensive use. A permanent flood area in lower Awash forms Lake Yardi, covering about 6,000 hectares, and a permanent swamp above the lake covers 5,000 hectares. Lake Kaddabasa is another swampy lake covering 20,000 hectares or more. These wetlands contain papyrus, *Typha*, and *Phragmites*, along with a variety of aquatic floodplain grasses and floating and submerged aquatics. Lake Abe, where the Awash River terminates, has wetland vegetation, including *Phragmites mauritianus*, *Cyperus papyrus*, and *Typha domingensis*.[178]

In the southwest, in Gambela regional state, vast areas, from the town of Jikawo in the north to Gog in the southeast to Gambela in the east, are wetlands. Flood plains and permanent wetlands also occur on the Akobo River along the Ethiopian-South Sudan border. There is an extensive swamp on the convergence of the Akobo town and Pibor River. There are also other permanent wetlands along the Gilo River.[179] The wetlands have diverse vegetation: tall grasses including asendabo (*Echinochloa* spp.) and sar (*Panicum maximum*), shrubs, and woody herbs. Dense small trees, acacia bushes, shenbeko (*Arundodonex*), shenkorageda (*Saccharum officinalis*), and temba (*Pennisetum petiolare*) cover the bank of the rivers, including the Baro.[180] Most of these wetlands are part of the Gambela National Park, but the recent introduction of commercial farming activities and resettlement programmes are threatening the integrity of these wetlands.

The Ethiopian government has no policy specifically for wetlands, so none has been designated for protection. The only formal government policies that mention wetlands are the Conservation Strategy of Ethiopia and the Water Resources Management Policy. Even these two policies only address wetlands indirectly. In the Conservation Strategy of Ethiopia, wetlands are addressed as regulators of water and water quality. In the Water Resources Management Policies, they are addressed for their biodiversity and role in reducing pollution. Wetlands are not part of the government policy at the national level.

Conclusion

Development activities have varying degrees of negative environmental impacts, particularly on water resources and related ecosystems. Ethiopia has no systematic water quality assessment program. Reports and studies on the topic are few; however, there are indications that water pollution is an increasing problem in some parts of the country. The population is growing along with industrialisation, urbanisation, and intensified agricultural and development activities. As a result, liquid, solid, and atmospheric pollutants are increasing. This will cause the quality of unprotected surface and groundwater to decline. Increased fertiliser, pesticide, and organic chemical use will contaminate water sources.

Clearly, Ethiopia must protect its surface and ground waters from being fouled and misused. There is also a need for conservation and efficient use of water resources in the country before they become scarce resources. Management of watersheds through extensive soil conservation, treatment of catchment area, protection of existing forests and increasing forest cover and the building of check-dams should be advanced.[181] It is also imperative that freshwater sources be protected from contaminants. Freshwater should be priced in relation to its value to health and production as well.

Notes

1 John Waterbury. 2002. *The Nile Basin: National Determinants of Collective Action*, New Haven: Yale University Press, p. 16.
2 Ministry of Water Resources (MoWR). 1998a. *Water Supply Development and Rehabilitation: Tariff and Asset Valuation Study*. Main Report, Addis Ababa: MoWR, p. 43.
3 Alan B. Dixon. 2003. *Indigenous Management of Wetlands: Experiences in Ethiopia*, London: Ashgate Publishing, p. 53.
4 British Geological Survey (BGS). 2001. *Groundwater Quality: Ethiopia, British Geological Survey*. Available online at www.wateraid.org/documents/plugin_documents/ethiopiagw.pdf (accessed 29 November 2010).
5 MoWR. 2002b. *Water Sector Development Program*. Final Report, Volume II, Addis Ababa: MoWR.
6 R. Johnston and M. McCartney. 2010. *Inventory of water storage types in the Blue Nile and Volta River Basins*. Colombo, Sri Lanka: International Water Management Institute, IWMI Working Paper 140, p. 7; BGS, p. 2.

7 Transitional Government of Ethiopia (TGE). 1994. *A Review of the Water Resources of Ethiopia*, Addis Ababa, p. 79.

8 Ibid., p. 84.

9 MoWR. 1998c. *Federal Water Policy and Strategy: Comprehensive and Integrated Water Resources Management*, Report, Vol. 1. Addis Ababa: MoWR, p. 15.

10 TGE, p. 87.

11 MoWR. 1998c, p. 15.

12 Tamiru Alemayehu. 2006. *Ground Water Occurrence in Ethiopia*, Addis Ababa: UNESCO, p. 75. Availabe online at www.eah.org.et/docs/Ethiopian%20groundwater-Tamiru.pdf (accessed 29 November 2010).

13 TGE, p. 89.

14 BGE, p. 5.

15 Ibid., p. 4. The ground waters in the Rift Valley are saline and high in fluoride content. Fluoride is ubiquitous in the Earth's crust and, thus, gets into groundwater through natural processes. Consumption of water with excessive fluoride is dangerous to human health. Most Ethiopians drink water with fluoride concentration of over 1.5 mg/l, a concentration high enough to cause dental fluorosis. A concentration of less than 1.5 mg/l provides protection against dental caries (Tamiru Alemayehu, p. 81).

16 BGS, p. 5.

17 G. M. Di Paola and Getahun Demissie. 1979. Geothermal energy: an inexhaustible resource of great economic importance for Ethiopia. *SINET: Ethiopian Journal of Science* 2, 2: 87–110.

18 Tamiru Alemayehu, pp. 69–70.

19 H. P. Huffnagel. 1961. *Agriculture in Ethiopia*, Rome: Food and Agriculture Organization, 1961, p. 53.

20 A. Swain. 1997. Ethiopia, the Sudan and Egypt: the Nile River dispute. *The Journal of Modern African Studies* 35, 4: 675–94.

21 Bo Appelgren, Wulf Klohn, and Undala Alam. 2000. *Water and Agriculture in the Nile Basin, Nile Basin*. Initiative Report to ICCCON, Food and Agriculture Organization, Rome, p. 3.

22 MoWR. 1997d. *Abbay River Basin Integrated Development Master Plan Project*, Volume V: Environment, Addis Ababa: MoWR, p. 41.

23 Ethiopian Valleys Development Studies Authority (EVDSA). 1992a. *A Review of Water Resources of Ethiopia*, Addis Ababa: EVDSA, p. 38.

24 Y. Arsano and I. Tamrat. 2005. Ethiopia and the Eastern Nile Basin. *Aquatic Sciences* 67: 15–27.

25 Food and Agriculture Organization of the United Nations (FAO). 2005. *Ethiopia: Irrigation in Africa in Figures – Aquastat Survey 2005*, Rome: FAO.

26 D. Jovanovic. 1985. Ethiopian interests in the division of the Nile River waters. *Water International* 10, p. 84; pp. 82–5.

27 MoWR. 1997d, pp. 1.5–1.7.

28 Dale Whittington and Elizabeth M. McClelland. 1992. Opportunities for regional and international cooperation in the Nile Basin. *Water International* 17: 144–54.

29 United Nations Educational, Scientific, and Cultural Organization (UNESCO). 2004. *National Water Development Report for Ethiopia*, Addis Ababa: World Water Assessment Program, p. 227. Available online at http://unesdoc.unesco.org/images/0014/001459/145926e.pdf (accessed 2 December 2010).

30 Georgi Galperin. 1978. *Ethiopia: Population, Resources, Economy*, Moscow: Progress Publishers, pp. 35–6.

31 Huffnagel, p. 52.

32 MoWR. 1996e. *Tekeze River Basin Integrated Development Master Plan Project: Volume I*, Main Report, Addis Ababa: MoWR.

33 EVDSA. 1989b. *Master Plan for the Development of Surface Water Resources in the Awash Basin*, Volume 4, Annex A: Climate and Hydrology, Addis Ababa: EVDSA, pp. 1–3.

34 Michele L. Thiene, Robin Abell, Melanie L. J. Stiassny, and Paul Skelton. 2005. *Freshwater Ecoregions in Africa and Madagascar: A Conservation Assessment*, Washington, DC: World Wildlife Fund, p. 353.

35 EVDSA (1992a), p. 64a; FAO. 1997. *Irrigation Potential in Africa: A Basin Approach*, Rome: FAO, p. 67.

36 FAO/Land and Water Development Division. 1997. *Irrigation Potential in Africa: A Basin Approach*, Rome: FAO, p. 67.

37 EVDSA. 1992a, p. 61.

38 Thiene et al., p. 353.

39 United Nations Economic Commission for Africa. 2000. *Trans-boundary River/Lake Basin Water Development in Africa: Prospects, Problems, and Achievements*, Addis Ababa: UNECA. p. 28.

40 EVDSA. 1992a, p. 57. The Omo is navigable in its last 100 kilometres.

41 Silas Mnyiri Mutia. 2009. Kenya's experience in managing climate change and water resource conflicts. In: *Climate Change and Trans-boundary Water Resources in Africa*, Workshop Report, 29–30 September, Mombasa, Kenya, p. 28.

42 Peter Greste. 2009. The dam that divides Ethiopians. *BBC News*, 26 March 2009. Available online at http://news.bbc.co.uk/2/hi/africa/7959444.stm (accessed 2 December 2010).

43 Terri Hathaway. 2008. What cost Ethiopia's dam boom? *International Rivers* (February), pp. 4–26; p. 17.

44 Indigenous peoples in Ethiopia's Lower Omo Valley include: Bodi, Daasanach, Kara (Karo), Muguji (Kwegu), Mursi, Nyangatom, Bashada, Bodi, Hamar, and Nyangatom, Arbore, Ari, Atse, Banna, Basketo, Birale (Ongota), Bodi, Daasanach (Galeb), Dime, Hamar, Kara (Karo), Maale, Muguji (Kwegu), Murile, Mursi, Nyangatom (Bume), Tsamai, Tsemako. *International Rivers*, Ethiopia's Gibe III Dam: Sowing Hunger and Conflict, Berkeley, CA, 2009.

45 African Resource Working Group. 2008. Environmental and social impacts of the proposed Gibe III hydroelectric project in Ethiopia's lower Omo River Basin. Available online at www.forestpeoples.org/sites/fpp/files/publication/2010/08/ethiopiahydroelecimpactsmay08eng.pdf (accessed 20 August 2012).

46 G. W. Coulter, B. R. Allanson, et al. 1986. Unique qualities and special problems of the African Great Lakes. *Environmental Biology of Fishes* 17, 3: 161–83.

47 L. C. Beadle. 1981. *The Inland Waters of Tropical Africa*, London: Longman Group Limited. Cited in Emily Peck, *Lake Turkana Conservation Science Program-WWF-US*. Available online at www.feow.org/ecoregion_details.php?eco=530 (accessed 7 December 2010).

48 J. D. Halfman and T. C. Johnson. 1988. High resolution record of cyclic climate change during the past 4 ka from Lake Turkana, Kenya, *Geology*, 16, 496–500; Cited in N. M. Velpuri and G. B. Senay. 2012. Assessing the potential hydrological impact of the Gibe III Dam on Lake Turkana water level using multi-source satellite data. *Hydrology and Earth System Sciences* 9: 2987–3027.

49 Thomas C. Johnson and John O. Malala. 2009. Lake Turkana and its link to the Nile. *Monographiae Biologicae* 89, 111: 287–304, p. 1.

50 Thieme et al., p. 152.

51 Emily Peck, *Lake Turkana Conservation Science Program-WWF-US*. Available online at www.feow.org/ecoregion_details.php?eco=530 (accessed 7 December 2010).

52 M. Fitzgerald. 1981. Sibiloi: The remotest park in Kenya. *Africana* 8, 4: 22; Kenya Wildlife Service. 2001. *Nomination Form for the Great Rift Valley Ecosystems Sites. Extension of Lake Turkana: South Island National Park*, Nairobi. Available online at www.unep-wcmc.org/sites/wh/pdf/Lake%20Turkana.pdf (accessed 7 December 2010).

53 L. C. Beadle. 1981. *The Inland Waters of Tropical Africa*, London: Longman Group Limited; A. J. Hopson. 1982. *Lake Turkana. A Report on the Findings of the Lake Turkana Project 1972–1975*, Volumes 1–6, London, UK: Overseas Development Administration; C. Leveque. 1997, *Biodiversity Dynamics and Conservation: The Freshwater Fish of Tropical Africa*, Cambridge, UK: Cambridge University Press.

54 Turkana mud turtle and three species of frogs that are endemic to the lake, including *Bufo chappuisi, B. turkanae*, and *Phrynobatrachus zavattarii*. R. H. Hughes and J. S. Hughes. 1992. *A Directory of African Wetlands*, Gland, Switzerland, Nairobi, Kenya, and Cambridge, UK: IUCN and UNEP.

55 Peter Greste. 2009. The dam that divides Ethiopians. *BBC News*, 26 March. Available online at http://news.bbc.co.uk/2/hi/africa/7959444.stm (accessed 2 December 2010).

56 Fitsum Merid. 2008. *National Nile Basin Water Quality Monitoring Baseline Report for Ethiopia*, Nile Basin Initiative, Trans-boundary Environmental Action Project, Addis Ababa, p. 15. Available online at http://nilerak.hatfieldgroup.com/English/NRAK/Resources/Document_centre/WQ_Baseline_report_Ethiopia.pdf (accessed 24 March 2011).

57 Galperin, p. 36.

58 Huffnagel, p. 53; EVDSA (1992a), p. 67.

59 World Meteorological Organization. 2003. *The Associated Program on Flood Management Integrated Flood Management: Case of Ethiopia*. Integrated Flood Management (by Kefyalew Achamyeleh). Technical Support Unit (ed.). Available online at www.apfm.info/pdf/case_studies/cs_ethiopia.pdf (accessed 11 March 2011).

60 Ethiopian Wildlife and Natural History Society (EWNHS). 1996. *Important Bird Areas of Ethiopia*, Addis Ababa: EWNHS and Birdlife International, p. 111.

61 Ethiopia's Rush to Build Mega Dams Sparks Protests. *The Guardian* (UK), 25 March 2010.

62 Tony Binns, Alan Dixon, and Etienne Nel. 2012. *Africa: Diversity and Development*, New York, Routledge, p. 94.

63 UNESCO, pp. 55–9.

64 MoWR. 2002b, p. 61.

65 World Bank. 2006. *Managing Water Resources to Maximize Sustainable Growth, Ethiopia*, Report 36000-ET, World Bank: Washington, p. 28.

66 National Research Council. 2009. *Emerging Technologies to Benefit Farmers in Sub-Saharan Africa and South Asia*, Washington, DC: The National Academies Press, p. 212.

67 Ethiopia Electric Power Corporation. 2009. Available online at www.eepco.gov.et (accessed 20 December 2009); UNESCO, p. 59.

68 Katrina Manson. 2014. Ethiopia uses electricity exports to drive ambition as an African power hub. *The Financial Times*, 16 February 2014. Available online at www.alloexpat .com/ethiopia_expat_forum/ethiopia-uses-electricity-exports-to-drive-ambition-as-an-a-t18916.html (accessed 10 July 2016).

69 Ethiopia Electric Power Corporation. 2009. World Bank, pp. 29–30. Available online at www.eepco.gov.et (accessed 20 December 2009).

70 Xan Rice. 2010. Green concerns over Ethiopia's dam plans. *The Sydney Morning Star*, 12 April.

71 Ethiopia's rush to build mega dams sparks protests. *The Guardian* (UK), 25 March 2010.

72 Ibid., pp. 77–8.

73 A. E. Cascao. 2008. Ethiopia – Challenges to Egyptian hegemony in the Nile Basin. *Water Policy* 10: 13–28; Yacob Arsano and I. Tamrat. 2005. Ethiopia and the eastern Nile Basin. *Aquatic Sciences* 67: 15–27; P. Kagwanja. 2007. Calming the waters: the east African community and the conflict over the Nile resources. *Journal of Eastern African Studies* 1, 3: 321–37; D. Whittington, X. Wu, et al. 2005. Water resources

management in the Nile Basin. The economic value of cooperation. *Water Policy* 7: 227–52; D. R. Karyabwite. 2000. *Water Sharing in the Nile River Valley*, UNEP; A. F. Metawie, 2004. History of Co-operation in the Nile Basin. *Water Resources Development* 20, 1: 47–63.

74 Ibid., pp. 78–9.

75 *Africa Report*, Ethiopia raises US$350 million for mega dam, Africa Report, 13 September 2011. Available online at http://nazret.com/blog/index.php/2011/09/13/ethiopia-raises-us-350-million-for-mega-dam (accessed 13 September 20--).

76 *Zemenawi Gibrnachen*, vol. 3, No. 1, Yekatit 2006 E.C. (The Ministry of Agriculture Bulletin), p. 14.

77 *ERATA News*, 2 April 2011. Ethiopia lays foundation for Africa's largest dam. Available online at www.ertagov.com/erta/erta-news-archive/38-erta-tv-hot-news-addis-ababa-ethiopia/574-ethiopia-lays-foundation-for-africas-biggest-dam.html (accessed 1 September 2012).

78 Full text of 'Declaration of Principles' signed by Egypt, Sudan and Ethiopia. http://nazret.com/blog/index.php/2015/03/30/full-text-of-declaration-of. Accessed 5 April 2015.

79 N. Stern. 2007. *Stern Review on the Economics of Climate Change*, London: Cambridge University Press. Available online at www.hm-treasurey.gov.uk/independent_reviews/stern_review_economics_climate_change/stern_review_report.cfm (accessed -------).

80 Theodore Panayotou. 1993. *Green Markets: The Economics of Sustainable Development*, San Francisco: Institute for Contemporary Studies, p. 80.

81 Fresh water lakes include: Abaya, Abijata, Ashange, Hawassa, Chamo, Hayik, Langano, Shalla, Tana, Ziway. Saline lakes are: Abe, Afambo, Afrera, Assela, Beseka, Chew Bahir, Gamari, Gargori, and Turkana. Crater Lakes are Bishoftu, Hora, Wonchi, and Zequala. Swamps are several, including: Chomen Dabus, Fogera Shallo, Dillu, Borkena, and Gedebassa. There also are swamps along rivers, including: Abbay, Alwero, Akobo, Baro, Gilo, and Wabi Shebelle. MoWR (1998c), pp. 11–3.

82 Ib Friis, Sebsebe Demissew, and Paulo van Breugel. 2011. *Atlas of the Potential Vegetation of Ethiopia*, Addis Ababa: Addis Ababa University Press, p. 14.

83 Huffnagel, p. 52; Galperin, p. 40.

84 EWNHS, p. 86.

85 Fitsum Merid, p. 12.

86 Galperin, p. 40.

87 Centre for Environmental Economics and Policy in Africa (CEEPA). 2006. *Climate Change and African Agriculture: Assessing the Impact of Climate Change on the Water Resources of the Lake Tana Sub-basin Using the Watbal Model*, Policy Note No.30, CEEPA: University of Pretoria, South Africa, p. 2. Available at www.ceepa.co.za/docs/POLICY%20NOTE%2030.pdf (accessed 8 December 2010).

88 MoWR. 1997d, p. 61.

89 EWNHS, pp. 89–90.

90 Amare Haileslassie, Fitsum Hagos, Everisto Mapedza, Claudia Sadoff, Seleshi Bekele Awulachew, Solomon Gebreselassie and Don Peden. 2008. *Institutional Settings and Livelihood Strategies in the Blue Nile Basin: Implications for Upstream/Downstream Linkages*, Working Paper 132, International Livestock Research Institute (ILRI) and International Water Management Institute, p. 8. Available online at www.indiaenvironmentportal.org.in/files/WOR132.pdf (accessed 29 November 2010).

91 Dagnew Mequanent and Aboytu Sisay. 2015. Wetland Potential, Current Situation and its Threats in Tana Sub-Basin, Ethiopia. *World Journal of Environmental and Agricultural Sciences*, 1, 1, p. 10.

92 Raymond F. Dasmann, John P. Milton, and Peter H. Freeman. 1973. *Ecological Principles for Economic Development*, New York: John Wiley, p. 204.

93 Berhanu Teshale, Ralph Lee, and Girma Zawdie. 2002. Development initiatives and challenges for sustainable resource management and livelihood in the Lake Tana region of northern Ethiopia. *International Journal of Technology Management and Sustainable Development* 1, 2: 111–24; pp. 15–6.

94 Thieme, et al., p. 150.

95 EWNHS, p. 164.

96 Ibid., p. 165.

97 Girma Kebbede. 2004. *Living with Urban Environmental Health Risks: The Case of Ethiopia*, London: Ashgate Publications, pp. 156–7.

98 Ethiopian Science and Technology Commission. 2004. *Proposal for Organizing a National Roundtable on Sustainable Consumption and Production in Akaki River Basin*, Addis Ababa. Available online at www.gadaa.com/AkakiRiverRoundtable.pdf (accessed 22 December 2010).

99 Office of Environmental Health Hazard Assessment, State of California. 2009. *Microcystis: Toxic Blue-Green Algae*. Available online at http://oehha.ca.gov/ecotox/pdf/microfactsheet122408.pdf (accessed 23 December 2010).

100 *BBC Video*, The Destruction of Ethiopia's Once Beautiful Lake Koka, 21 February 2009. Available online at http://nazret.com/blog/index.php?title=the_destruction_of_ethiopia_s_once_beaut_1&more=1&c=1&tb=1&pb=1 (accessed 21 December 2009).

101 Kiran Khalid, Akaki River. Available online at www.gadaa.com/AkakiRiver.html (accessed 21 December 2010).

102 P.C. Prabu. 2009. Impact of heavy metal contamination of Akaki River of Ethiopia on soil and metal toxicity on cultivated vegetable crops. *Electronic Journal of Environmental, Agricultural and Food Chemistry* 8, 9: 818–27. Available online at http://ejeafche.uvigo.es/component/option,com_docman/task,doc_view/gid,534/ (accessed 21 December 2010).

103 Solomon Kibret, Matthew McCartney, Jonathan Lautze, and Gayathree Jayasinghe. 2009. *Malaria Transmission in the Vicinity of Impounded Water: Evidence from the Koka Reservoir, Ethiopia*, IWMI Research Report 132, Colombo, Sri Lanka: International Water Management Institute, 2009. Available online at www.iwmi. cgiar.org/Publications/IWMI_Research_Reports/PDF/PUB132/RR132.pdf (accessed 22 December 2010).

104 UNESCO, p. 66.

105 EWNHS, p. 164; MoWR. 2002b, p. 85.

106 MoWR. 1998c, p. 30.

107 CSA. 2010. *Statistical Abstract 2009*, Addis Ababa: CS, p. 5.

108 E. Gilbert, Y. Travi, M. Massault, T. Chernet, F. Barbecot, and F. Laggoun-Defarge, 1999. Comparing carbonate and organic AMS-14C ages in Lake Abiyata sediments (Ethiopia). *Hydrochemistry and Paleoenvironmental Implications Radiocarbon* 41: 271; Cited in HuibHengsdijk and Herco Jansen. 2006. *Agricultural Development in the Central Ethiopian Rift Valley: A Desk-Study on Water-Related Issues and Knowledge to Support a Policy Dialogue*, Plant Research International B.V., Wageningen.

109 M. Bolton. 1969. Rift Valley Ecological Survey, UNESCO, p. 14; Emil K. Urban. 1970. Ecology of water birds of four Rift Valley lakes in Ethiopia. *Ostrich Sup.*, 8: 315–21; p. 318.

110 Herco Jansen, Huib Hengsdijk, Dagnachew Legesse, Tenalem Ayenew, Petra Hellegers, and Petra Spliethoff. 2007. *Land and Water Resources Assessment in the Ethiopian Central Rift Valley: Project: Ecosystems for Water, Food and Economic Development in the Ethiopian Central Rift Valley*, Alterra-rapport. Wageningen, Alterra, p. 10.

111 Dagnachew Legesse and Tenalem Ayenew. 2006. Effect of improper water and land resource utilization on the central Main Ethiopian Rift lakes. *Quaternary International* 148, 1: 8–18; p. 14.

112 Thieme et al., p. 182.

113 Tenalem Ayenew. 2002a, p. 98.

114 Ibid., p. 98.

115 Dagnachew Legesse and Tenalem Ayenew, p. 14.

116 EVDSA. 1992. *Institutional Support Project: A Review of the Water Resources of Ethiopia*, Report No. M2/C5/512/1. Addis Ababa: Richard Woodroofe and Associates, p. 21.

117 CSA. 2010. *Statistical Abstract 2009*, Addis Ababa: CSA, p. 5.

118 M. Bolton. 1969. *Rift Valley Ecological Survey*, UNESCO, p. 2.

119 Tenalem Ayenew (2002a), p. 87.

120 Geological Survey of Ethiopia, Current Exploration and Mining. Available online at www.telecom.net.et/~walta/profile/html/current_status.html (accessed 13 December 2010).

121 Ibid.

122 Tenalem Ayenew. 2002. Recent changes in the level of Lake Abiyata, Central Main Ethiopian rift. *Journal of Hydrological Science* 47, 3: 493–503. Available online at http://www.tandfonline.com/doi/abs/10.1080/02626660209492949 (accessed 10 December 2010).

123 Tenalem Ayenew. 2002a, p. 98.

124 CSA. 2010. *Statistical Abstract 2009*, Addis Ababa: CSA, p. 5.

125 Galperin, p. 39.

126 Bolton, p. 5.

127 TGE, p. 21.

128 CS. 2010. *Statistical Abstract 2009*, Addis Ababa: CSA, p. 5.

129 Dagnachew Legesse and Tenalem Ayenew, p. 8; R. B. Wood, M. V. Prosser, and F. M. Baxter, 1978. Optical characteristics of the Rift Valley lakes, Ethiopia. *SINET: Ethiopian Journal of Science* 1, 2: 73–85; p. 73.

130 Urban, p. 318; EWNHS, p. 167.

131 Including: White-breasted Cormorant, Great White Pelicans, Kittlit's Sandplover, Spur-winged Plover, White-winged Black Tern, and Lesser Flamingo. The area also attracts Little Rock Thrush and White-winged Cliff-chat. Scrub lures Singing Cisticola, Blue-breasted Bee-eater, Ruppell's Weaver, Lesser Masked Weavers, Grey-headed Gull, Common Black-headed Gull, Slender-billed Gull, Whiskered and White-winged Gull, Stout Cisticola, and Thrush Nightingale. EVDSA. 1992a, p. 40.

132 Ibid.

133 The section on Hawassa Lake benefits from the work of Zerihun Desta. 2000. Challenges and opportunities of Ethiopian wetlands: the case of Lake Awassa and its feeders. In: *Proceedings of a Seminar on the Resources and Status of Ethiopia's Wetlands*, Yilma D. Abebe and Kim Geheb (eds.), The World Conservation Union (IUCN), pp. 67–74.

134 CSA. 2010. *Statistical Abstract 2009*, Addis Ababa: CSA, p. 5.

135 Bolton, p. 11.

136 World Meteorological Organization. 2003. *The Associated Program on Flood Management Integrated Flood Management: Case of Ethiopia: Integrated Flood Management*, Technical Support Unit, Kefyalew Achamyeleh (ed.). Available online at www.apfm.info/pdf/case_studies/cs_ethiopia.pdf (accessed 11 March 2011).

137 The most notable of these include the African pygmy goose, white-backed duck, marabou stork, silvery-cheek hornbill, grey kestrel, African fish eagle, red-billed teal, black crake, yellow-billed egret, yellow-fronted parrot, blue-headed coucal, Bruce's green pigeon, white-rumped babbler, storks, terns, plovers, hooded vulture, woodland kingfisher, red-headed weaver, and the endemic black-winged lovebird. Philip Briggs and Brian Platt. 2009. Ethiopia: *The Bradt Travel Guide*, Guilford, CT: The Globe Pequot Press, p. 465.

138 Zerihun Desta, pp. 67–74.

139 Ibid.

140 Francis L. Gordon. 2000. *Ethiopia, Eritrea and Djibouti*, London: Lonely Planet Publications, p. 229.

141 Friis et al., p. 136.

142 Galperin, p. 39.

143 Andrew S. Goudie. 1975. Former lake levels and climatic change in the Rift Valley of southern Ethiopia. *Geographical Journal* 141: 177–194. The lake has many islands, including Aruro, Gidicho, Galmaka, Alkali, and Welege.

144 CSA. 2010. *Statistical Abstract 2009*, Addis Ababa: CSA, p. 5.

145 Briggs and Blatt, pp. 514–5.

146 EWNHS, p. 209.

147 Galperin, p. 39.

148 Briggs and Blatt, p. 533; EWNHS, p. 209.

149 EWNHS, 1996, p. 210.

150 Including: near threatened Lesser Flamingos, Storks, Waterfowl and Palaearctic Waders, Gull-billed Tern, White-faced Whistling Duck, Fulvous Whistling Duck, Pink-breasted Lark, Scaly Chatterer, Sparrow Weaver, Shelley's Starling Parrot-billed Sparrow, and Grey-headed Silverbill. EWNHS, 1996, p. 210; Briggs and Blatt, p. 534.

151 Institute of Biodiversity. 2008. *Conservation, Ecosystems of Ethiopia*, Addis Ababa. Available online at www.ibc-et.org/ (accessed 6 January 2011).

152 UNESCO, p. 120.

153 Institute of Biodiversity. 2008. *Conservation, Ecosystems of Ethiopia*, Addis Ababa. Available online at www.ibc-et.org/ (accessed 6 January 2011).

154 R. Johnston, and M. McCartney. 2010. *Inventory of Water Storage Types in the Blue Nile and Volta River Basins*. Colombo, Sri Lanka, International Water Management Institute, IWMI Working Paper 140.

155 Shewaye Deribe. 2008. Wetland management aspects in Ethiopia: situation analysis. In: *The Proceedings of the National Stakeholders' Workshop on Creating National Commitment for Wetland Policy and Strategy Development in Ethiopia*, 7–8 August, Addis Ababa, Ethiopia, pp. 14–27. Available online at www.ewnra.org.et/files/Proceedings%20of%20the%20national%20stakeholders%27%20workshop.pdf (accessed 1 December 2010).

156 Jonathan McKee. 2007. *Ethiopia: Country Environmental Profile*, p. 33. Available online at http://ec.europa.eu/development/icenter/repository/Ethiopia-ENVIRONMENTAL-PROFILE-08-2007_en.pdf (accessed 10 April 2011).

157 Institute of Biodiversity. 2008. *Conservation, Ecosystems of Ethiopia*, Addis Ababa. Available online at www.ibc-et.org/ (accessed 6 January 2011). Important wetlands within the country include Lake Tana and its associated wetlands (the Fogera floodplains); the Dembia floodplains; the Ashange and Hayk lakes; the wetlands of the Bale Mountains including alpine lakes, such as Garba Guracha and swamps and floodplains; the wetlands of the western highlands, such as Ghibe and Gojeb in Kaffa and Illu Ababor; the lakes of Bishoftu, including the crater Lakes Hora, Bishoftu, Guda, and Zukuala and the Green, Babugaya, and Bishoftu lakes; the lakes and wetlands in the southwest Great Rift Valley, such as Lakes Ziway, Langano, Abijata, Shala, Awassa, Chelekelaka, Abaya, Chamo, Chew-Bahir, and Turkana; the lakes and wetlands of the Awash River basin, such as Dillu Meda and Aba Samuel in the upper Awash; the Lake Beda sector, which includes the Gewane lakes and swamps, the Dubiti, Afambo, and Gemari lakes and swamps, and Lake Abe and the delta; the lakes of the Afar Depression, which are Lakes Afrera, Asale, and the Dallol Depression itself; the western river floodplains, such as the Alwero, Baro, Akobo, Gillo, Chomen, Fincha'a swamp, Dabus swamp, and the Beles floodplain; and the artificial wetlands and dams, such as the Koka, Fincha, Melka-Wakana, Gilgel Gibe,

and other hydropower dams and municipal and other reservoirs, such as dams, aquifers, and wells. UNESCO, pp. 122–3.

158 Ibid., p. 125.

159 Ibid., pp. 126–7.

160 Ibid., p. 127.

161 Dixon, p. 63.

162 Alan Dixon and Adrian Wood. 2001. *Sustainable Wetland Management for Food Security and Rural Livelihoods in South-west Ethiopia: The Interaction of Local Knowledge and Institutions, Government Policies and Globalization.* Paper Prepared for Presentation at the Seminaire sur L'amenagement des zones marecageuses du Rwanda, 5–8 June, National University of Rwanda and the University of Huddersfield and Wetland Action.

163 Ibid.

164 Adrian Wood. 2000. Wetlands, gender, and poverty: some elements in the development of sustainable and equitable wetland management. In: *Proceedings of a Seminar on the Resources and Status of Ethiopia's Wetlands*, Yilma D. Abebe and Kim Geheb (eds.), The World Conservation Union (IUCN).

165 EWNHS, pp. 163–4.

166 Thieme et al., p. 180.

167 E. J. Mwendera, M. A. Mohamed Saleem, and A. Dibabe. 1997. The effect of livestock grazing on surface runoff and soil erosion from sloping pasture lands in the Ethiopian highlands. *Australian Journal of Experimental Agriculture* 37: 421–30.

168 Messele Fisseha. 2003. Water resources policy and river basin development as related to wetlands, In: *Wetlands of Ethiopia. Proceedings of a Seminar on the Resources and Status of Ethiopia's Wetlands*, Yilma D. Abebe and Kim Geheb (eds.), Nairobi, IUCN Eastern Africa Regional Office, p. 78.

169 Briggs and Platt, p. 338.

170 Messele Fisseha. 2003. Water resources policy and river basin development as related to wetlands, In: *Wetlands of Ethiopia. Proceedings of a Seminar on the Resources and Status of Ethiopia's Wetlands*, Yilma D. Abebe and Kim Geheb (eds.), Nairobi, IUCN Eastern Africa Regional Office, p. 9; Sir William Halcrow et al. 1989. *Master Plan for the Development of Surface water Resources in the Awash Basin, Ethiopian Valleys Development Authority*, Vol. 8, Annex J: Irrigation and Drainage.

171 Degefa Tolossa and Axel Baudouin. 2004. Access to natural resources and conflicts between farmers and agro-pastoralists in Borkena Wetland, Northeastern Ethiopia. *Norwegian Journal of Geography* 58, 3: 97–112.

172 Ibid., p. 376.

173 Zerihun Desta. 2000. Challenges and opportunities of Ethiopian wetlands: the case of Lake Awassa and its feeders. In: *Proceedings of a Seminar on the Resources and Status of Ethiopia's Wetlands*, Yilma D. Abebe and Kim Geheb (eds.), The World Conservation Union (IUCN).

174 Irit Eguavoen. 2009. *The Acquisition of Water Storage Facilities in the Abbay River Basin, Ethiopia*, Working Paper Series 38, Center for Development Research (ZEF), University of Bonn.

175 Bezuayehu Tefera and Geert Sterk. 2008. Hydropower-induced land use change in Fincha'a watershed, Western Ethiopia: analysis and impacts. *Mountain Research and Development* 28, 1: 72–80.

176 Watershed, Ethiopia: Land Use Changes, Erosion Problems, and Soil and Water Conservation Adoption, Wageningen: Wageningen University, The Netherlands. No date. Available online at www.mekonginfo.org/mrc_en/doclib.nsf/0/e1dfbbefb9263 e6b4725724a00123f75/$file/07_abstr_sterk_tefera.pdf (accessed 5 January 2011).

177 EWNHS, pp. 159–61.
178 R. H. Hughes and J. S. Hughes. 1992. *A Directory of African Wetlands*, Nairobi: UNEP and IUCN, pp. 165–6.
179 Ibid., p. 169.
180 EWNHS, p. 111.
181 V. S. Ganesamurthy. 2011. *Environmental Status and Policy in India*, New Delhi: New Century Publications, p. xxv.

6 Wildlife and protected areas

The Ethiopian landscape has undergone remarkable modifications due to physical and chemical weathering caused by constant atmospheric exposures, volcanic activities, rainfall, wind, glacial erosion of the highly elevated areas, and biological activities. The result was the formation of diverse topographical conditions with extreme vertical range from over 4,620 metres (m) above sea level at Mount Dashen in the Semien Mountains to one of the lowest and hottest places on earth in the Dallol Depression in the Rift Valley, 120 m below sea level. Between these two contrasting elevations exists a series of contiguous bioclimatic stages ranging from the extreme desert with no vegetation of any kind to semi-desert-savanna with *Acacia panicum* to mountains with *Juniperus, Olea, Podocarpus*, and *Hagenia abyssinica*, and afro-alpine formations such as *Alchemilla* and *Lobelia*.[1] This geologic and geographic transformation has stimulated the evolution of a broad spectrum of plant and animal species, many of which are replicated nowhere else on earth. Human activities over the last five millennia, however, have profoundly transformed the landscape of northern Ethiopia. Woody vegetation, which once covered much of the highlands, has been cleared for agriculture and grazing. Mountain forests have been relentlessly removed for fuel and construction. The habitats of most wild creatures and plants have suffered severely in much of the highland areas of the country with a long history of human occupancy. Most of the original highland vegetation is long gone. Many of the species of wildlife in the highlands, especially the larger ones, have also been eliminated. Lions, elephants, tigers, and other large animals are said to have roamed the highlands of northern Ethiopia. Almost all of these animals have been decimated.[2]

Despite extensive modification of the Ethiopian landscape, the country still provides a natural home for many types of endemic (i.e., not found elsewhere) wildlife, avifauna, and plant species. The diversity and the endemism rate of fauna and flora in the country are remarkable. Ethiopia is one the top 25 mega-diverse countries in terms of high vertebrate species (mammals, birds, reptiles, and amphibians) and endemic languages.[3] This is perhaps the result of millions of years of isolation of the country's extensive highland ecosystems and towering mountain ranges from surrounding lowland regions. Of the more than 7,000 flowering plant species that have thus far been classified in the country, some 800 (11.4 per cent) are endemic species.[4] Over 118 crop species and their wild relatives have been identified. There are 277 species of mammals, 35 of which are found exclusively in Ethiopia.[5] There are also 150 species of freshwater fish, 201 species of reptiles, and 63 species of amphibians. Of these, four freshwater fish, ten reptiles, and 34 amphibians are found only in Ethiopia.[6]

Ethiopia is also very rich in avifauna diversity. The country provides habitat for about 862 species of birds, of which 28 are found nowhere else in the world. These numbers rank Ethiopia as one of the important centres of avifauna in Africa. The central and southern highlands and the Juba-Shebelle River valleys are among the 63 globally recognised endemic bird areas. The country also provides breeding and resting sites for a large number of European and central Asian migratory birds during the northern winter season. A few globally threatened bird species make regular visits to 50 bird sites in the country. There is, however, serious concern regarding the erosion of avifauna diversity. Two hundred sixty-one bird species (30 per cent) are of international concern. Of these, 31 bird species are listed as globally threatened species: 5 endangered, 12 vulnerable, and 14 near-threatened species.[7]

This chapter examines the status and functions of national parks and other protected areas in Ethiopia, focusing mainly on two aspects: first, the problems associated with the management of protected areas; and second, the impacts of national parks and other protected areas on the livelihood systems of local communities.

Wildlife and protected area management

Ethiopia's early public initiative concerning wildlife protection was recorded in the first decade of the last century. In 1908, Emperor Menelik II issued a proclamation, which prohibited large-game hunting, including elephants.[8] However, Menelik's expressed wishes went largely unheeded due to the lack of an enforcement mechanism. In 1944, further laws were passed by Emperor Haile Selassie, primarily aimed at regulating hunting of certain wildlife species on royal game grounds, which are now the Awash National Park.[9] Although concern for the protection of wildlife was expressed in the legislation and a few protected areas were set up, his government never adequately protected wildlife. Throughout the 1940s and 1950s, trophy hunting by both nationals and expatriates persisted. The lack of effective laws to control illegal hunting and the widespread availability of modern firearms resulted in the demise of large quantities of big games for their meat, skins, and ivory. Skins of protected animals, notably the Colobus monkey and leopard, were traded in open markets in the capital as well as in other major regional urban centres.[10]

The United Nations established the principle of nature conservation in the 1950s and urged member governments to set up parks to conserve some unique ecosystems. In Ethiopia, the idea of creating national parks and other protected areas emerged in the late 1950s. Two principal factors may have provided the impetus.

First, the decade of the 1950s saw an increased onslaught of large animals and the destruction of their habitats because of poaching, unregulated hunting, deforestation, and clearance of land for cultivation. Second, the government also perceived the possibility of generating foreign exchange earnings through tourism by preserving the country's rich biological diversity, scenic beauty, and natural wonders. Consequently, during the early 1960s, the foundations for establishing and managing unique ecosystems were laid when the government solicited international support to create national parks and wildlife conservation areas. Nature conservation experts from United Nations Educational, Scientific and Cultural Organization (UNESCO) and United States-based conservation organisations made a series of field investigations in the country to designate areas of high conservation potential for national parks and wildlife reserves.[11] Such areas were identified primarily on the basis of the distribution and abundance of larger wildlife species. The result was the establishment of a number of national parks and other types of protected areas in various parts of the country in the late 1960s, including the Awash and the Semien Mountains National Parks. With financial support and technical assistance from the World Wildlife Fund, the USAID and other expatriate agencies, the Ethiopian Wildlife Conservation Organization (EWCO) was created within the Ministry of Agriculture as the managing and regulatory agency for the country's protected areas.[12] The EWCO's main responsibilities at the time included: (1) the conservation of wildlife, particularly endangered species; (2) the protection of wildlife habitats and areas of ecological significance; (3) the establishment of conservation areas in the form of national parks, wildlife sanctuaries, and reserves; (4) the control of wildlife utilisation and products; and (5) the creation of awareness toward wildlife and conservation through education, training, workshops, and research.[13]

The government enacted a series of laws in 1970, 1972, 1974, and 1980 to strengthen existing management and utilisation regulations as well as to extend protections to additional wildlife habitats in various regions of the country. In 1995, by the federal government's policy of devolving state power, management of all protected areas was transferred to the newly established regional state governments. However, the transfer from the central government to regional state control has not been fully realised. As a result, the EWCO carries much of the responsibility for managing protected areas of the country by providing financial and material resources, wardens, training, and workshops to regional states in an effort to build their human resource capacities to establish and manage wildlife resources within their jurisdiction.

Initially, national parks were the principal form of protected areas in the country. Over the years, however, other types of protected areas were added to nature reserves. Today, Ethiopia has over 60 national parks, sanctuaries, wildlife reserves, and controlled hunting areas (Table 6.1), but quite a few of these areas are well managed. In all, they cover an area equal to 16.4 per cent of the total land area of the country, significantly higher than the sub-Saharan African average (10.9 per cent) and the world average (10.8 per cent).[14] Twenty-eight of the

Table 6.1 National parks and protected areas in Ethiopia

Names	Size (hectares)
Abijata-Shala Lakes	88,700
Alatish	266,600
Awash	75,600
Bale Mountains	247,100
Gambela	506,100
Genale	385,800
NechSar	51,400
Omo	406,800
Semien Mountains	41,200
Yangudi-Rassa	473,100
Chebera Churchura	119,000
Dati Wole	143,100
Denkoro Chaka	38,117
Gibe Sheleko	24,800
Kuni-Muktar	150,000
Mago	194,200
Maze	20,200
Yabelo	250,000
Senkelle Swayne's Hartebeest Sanctuary	5,400
Alledeghi	193,389
Chelbi	421,200
Mile Serdo	650,354

James Young. 2012. Ethiopian Protected Areas: A 'Snapshot'. Available online at http://phe-ethiopia. org/admin/uploads/attachment-1167-Eth%20Protected%20Areas%20Snapshot%201_4_2012.pdf (accessed 8 August 2014).

protected areas cover 1,000 square kilometres or more, while five protected areas measure more than 10,000 square kilometres each.[15] However, only two of these protected areas (the Awash and the Semien Mountains National Parks) were ever legally constituted. While the primary purpose of all protected areas is to preserve wildlife species, the level of protection varies.

The National Parks are set aside for the purpose of preserving and protecting wildlife and scenic beauty, as well as to advance scientific, educational, and recreational interests. Prohibited activities include human settlements, cultivation, grazing cattle, hunting, burning vegetation, and the extraction of natural resources, such as fuelwood and timber. Non-extractive activities (for example, hiking and safari viewing) are allowed to support the development and management of the parks. Up until 2000, there were nine national parks covering an estimated 23,000 square kilometres that include the Abijata-Lakes, Awash, Bale Mountains, Gambela, Mago, Nech Sar, Omo, Semien Mountains, and Yangudi-Rassa. Recent years have seen the establishment of additional parks, including Kafeta Sheraro (Tigray), Dati, Gibe Sheleko (Oromiya), Chebera Churchura, Maze, Loka Abaya (Southern Nations), Alatish, Borana Sayint, and Bahir Dar Blue Nile (Amhara), and Gerale (Somali).[16]

Wildlife Sanctuaries are the particular protection of species or habitat, which is endangered or threatened. Activities such as grazing cattle, establishing settlements, and hunting are prohibited in wildlife sanctuaries unless the state grants written permission. Wildlife sanctuaries cover about 10,000 square kilometres. Wildlife sanctuaries include Didessa, Babille, Kuni-Muktar, Senkelle, Stephanie, and Yabelo.

Wildlife Reserves are established to protect wildlife and their habitats, but partial exploitation, such as grazing cattle, settlement, and cultivation, is allowed with permission from the state. The possession of firearms or the hunting of animals, unless acting by the conditions of the game capture permit or with the written permission of the general manager or game warden, is prohibited. Wildlife reserves cover 30,000 square kilometres of land. Wildlife reserves include Alledeghi, Awash West, Bale, Chew Bahir, Erer, Gewane, Mille-Serdo, Shire, and Tama.

Controlled Hunting Areas cover an estimated 140,000 square kilometres. They are usually adjoining national parks or wildlife sanctuaries that serve as buffer zones that balance the protection and the limited exploitation of wildlife species. Prohibited activities include grazing of cattle, settlements, and killing animals unless acting by the conditions of a permit,[17] but the ban has been imposed since 1994 to enable stock taking. Controlled hunting areas include Afdem-Gewane, Akobo, Arsi, Awash West, Bale, Borana, Boyo, Chercher-Arba Gugu, Dabus, Eregota, Hararghe-Wabi Shebelle, Jikawo, Lower Wabi Shebelle, Maze, Mizan Teferi, Murle, Omo West, Segen, and Tado.[18] The following chapter examines the state of some of the oldest national parks and wildlife sanctuaries in the country and the controversies and management problems surrounding their protection.

The Awash National Park

Established in 1966, the Awash National Park is the oldest park in Ethiopia. It is in the Afar regional state, about 225 kilometres east of Addis Ababa. The park includes an old imperial hunting reserve. It received its legal status in 1969.

The park is one of eight conservation areas within the Awash River basin in the Ethiopian Rift Valley. The park is on the left bank of the Awash River, which swings north immediately after leaving the park. The park covers an area of approximately 827 square kilometres, most of which lies at an altitude of about 900 m. In the middle of the park is Mount Fantale, a dormant volcano, reaching a height of slightly over 2,000 m above sea level. Much of the area is scorching. The average annual temperature exceeds 35 degrees centigrade. Rains are scarce and irregular, 200 millimetres (mm) or less per year in the north, and rarely exceeding 500 mm per year in the south. Scattered short acacia trees, small bushy plants, and grass species cover the plains. A narrow belt of larger and denser riverine vegetation, including acacia, tamarind, and fig trees, is found along the river.

The Awash National Park provides home for a variety of large mammals, including Beisa Oryx, greater and lesser kudus, Soemmering's gazelles, Grevy's zebras, Swayne's hartebeests, Salt's dik-dik, Anubis and Hamadryads baboons, vervet and Colobus monkeys, mountain reedbucks, klipspringers, and wild cats.[19] Some animals have disappeared altogether from the area. For example, elephants, buffalos, lions, and leopards, which roamed the valley only half a century ago, no longer inhabit the park. Crocodiles and hippopotamuses make a home in the Awash River. Birdlife abounds, with over 460 species identified thus far.[20]

The Awash National Park has had management problems right from its establishment. The park was created on lands traditionally used by the Afar, Kerrey, Ittu, and Argoba transhumant pastoralists. The park poses an enormous problem for these pastoral communities, as it prohibits access to dry season grazing lands and water points. For these communities, access to the Awash River and its associated floodplains, swamps, and marshes is critical to the successful sustenance of the long-established and ecologically well-adapted extensive grazing system.[21] Prior to the establishment of the park, the pastoral communities had already lost much of their grazing land to large irrigation schemes and agriculture. Approximately 70,000 hectares of previously dry season grazing land was under irrigation in the upper, middle, and lower Awash Valley. It amounts to 43 per cent of the total irrigated agricultural areas in the country.[22] Most of these areas have been developed since the early 1960s after the construction of the Koka Dam regulated the flow of the Awash River.

The creation of the Awash National Park thus led to sharp conflicts with the local communities attempting to maintain their customary rights to grazing land and water in the valley. From the outset, the government imposed severe restrictions on local grazing to maintain effective control over the park. Pastoralists who grazed their cattle inside the park were routinely expelled by force (as discussed in Chapter 3).

The park escaped the destruction that overcame other protected areas and wildlife reserves in the country following the collapse of the Derg Regime in

1991. Many local communities attacked neighbouring and nearby protected areas by looting and destroying infrastructure in the reserves. Acting out of retaliation, the people also illegally hunted animals.[23] But the Awash National Park remained safe from the attacks because the Wildlife Conservation Society, in conjunction with EWCO, started a community-based conservation project in 1990 to improve the indigenous peoples' attitudes toward the park and encouraged their cooperation with conservation efforts. The five-year management plans had been put in effect before the unrest and continued after the political transition.[24]

In the new park plan, EWCO acknowledged the ineffectiveness of its exclusionary policies and recognised the instrumentality of local peoples' involvement in managing the protected areas. EWCO's new park plan envisaged several undertakings, including allowing pastoralists to access pasture in the park in times of drought, working with government and non-government organisations to provide social services (particularly health, veterinary, and technical services to local communities), and educating park staff in community-based conservation and wildlife protection. Under the government's 1995 de-centralisation policies, management responsibilities of protected areas were transferred to regional states. However, EWCO retained control of the Awash National Park because the park lay in the two regional states, Oromiya and Afar. Partitioning of the park was considered impractical for management and risky for wildlife, yet the whole granting of autonomy and transfer of authority turned out to be somewhat irrelevant, as significant allocation of funds did not accompany the transfer of responsibilities.[25]

Figure 6.1 Awash National Park: Beisa Oryx.

Despite these changes, indigenous communities in and around the park find it increasingly difficult to survive because much of the land is designated as protected. While the new plan attempts to foster understanding and cooperation between local communities and EWCO, it does not promote ways for the indigenous people to live, use natural resources for their subsistence, and generate income.[26] Continued harvesting of fuelwood and timber, as well as increased competition between livestock and wildlife, have led to many animals being driven off and an increase in disease transmission. Wildlife is declining considerably (including the majestic Beisa Oryx shown in Figure 6.1), except in the case of species that favour shrubland habitats, such as Salt's dik-dik (*Madoquo saltiana*), lesser kudu (*Tragelaphus imberbis*), and common waterbuck (*Kobus ellipsiprymnus*).[27]

Yangudi-Rassa National Park

The 4,736-square kilometres Yangudi-Rassa National Park and its northern extension of the Mille-Serdo Wild Ass Reserve (8,766 square kilometres) are located almost entirely to the east of the Awash River. It is in the centre of the Afar regional state between the towns of Mille and Gewane, southwest of the Yangudi Mountain, which is surrounded by the Rassa Plains. The altitude of the area ranges from below sea level in the Danakil depression to 400 m in the Teo Plains to 1,400 m above sea level in the broken and eroded hills near the Awash River. Temperatures are extremely high and very little rain falls in this semi-desert area. The vegetation, therefore, consists of semi-arid grass and short trees with succulent scrub.

The park and the adjoining wild ass reserve were created in 1969 to protect the critically endangered African Wild Ass (*Equus africanus*), otherwise known as the Somali Wild Ass – *Equus asinus somalicus*, and other rare species. Other protected wildlife in the park includes Soemmering's gazelles, Beisaoryx, Gravy's zebras, and gerenuk and hamadryads baboons.[28] All in all, 36 species of mammals and 229 species of birds, of which two are endemic, have been identified in the park.[29] The park also plays an important role in safeguarding an archeological site of national and international interest. It was at the renowned site of Hadar where the oldest evidence of human origins was unearthed in the form of Lucy (also given Amharic name: *Dinkenesh*), a fossilised skeleton of a female hominid (*Australopithecus afarensis*) who walked this part of the earth more than three million years ago. A new hominid, *Ardi*, was also found here in 1994, pushing back the earliest known hominid date to 4.4 million years ago.[30] The Hadar area forms part of a 50-kilometre stretch in the complex of eroded hills, which skirt the eastern side of the Awash River, where archeological digs are ongoing.

The African Wild Ass also lives in neighbouring Eritrea and Somalia. The principal habitat of the African wild ass is arid, semi-arid bushland, and grassland. They are highly adaptive to harsh, dry environments, for they possess hardy digestive systems that can break down desert vegetation and extract moisture from food efficiently. Like camels, they can also survive without water for relatively long periods.[31] In Ethiopia, the flat plains between Dubti and Serdo in the Afar regional state are the home of the largest wild ass population in the region. Their favoured forage consists of *Aristida* sp., *Chrysopogon plumulosus, Dactyloctenium schindicum, Digiteria* sp., *Lasiurus scindicus*, and *Sporobolus iocladus*.[32] Lake Hertale, south of Gewane town, is a major source of water for the wild ass.

The protection of the park has been minimal since its establishment, mainly due to the presumption that the scorching and harsh environment would prohibit human interference. However, the African Wild Ass population in the park continues to diminish in number from competition for limited pasture and water sources with domestic animals, recurrent and devastating droughts, and interbreeding with the domestic donkey.[33] In 2007, there were fewer than 200 African Wild Ass in Ethiopia and 400 in Eritrea.[34] Wild asses are also hunted for food and medicinal purposes because their 'bodyparts and soup made from their bones are used for treating tuberculosis, constipation, rheumatism, backache, and bone-ache.'[35] The meagre vegetation in their environments forces them to live in isolation from one another, preventing them from reproducing.[36] Their restricted range and fragmented populations also make them highly vulnerable to human assault.

Semien Mountains National Park

The Semien Mountains National Park (SMNP) comprises 190 square kilometres, with a 400-square kilometre buffer zone around it. The park occupies a broad, undulating plateau of vast, grassy plains in the high mountain massif in the Amhara regional state in northern Ethiopia. Elevations range from 2,000 m to 4,620 m above sea level at Ras Dashen. Rainfall in the region increases with altitude. A maximum annual rainfall of 1,000 to 1,200 mm occurs at altitudes between 3,200 m and 4,500 m above sea level. Snow occasionally occurs above 3,800 m but does not form a permanent snow cover on the mountaintops. Low temperatures average between 9 degrees and 14 degrees centigrade annually, and frequent frosts make these high altitude areas unfavourable for the cultivation of most grains and pulses. Rainfall in the lower elevations (2,000–3,000 m) ranges between 1,000 mm and 1,500 mm. Annual temperatures average between 14 degrees and 18 degrees centigrade. This topographically less rugged altitudinal belt is favourable for the cultivation of most highland crops and pulses and, hence, much of this belt is heavily cultivated.

Much of the original natural vegetation of the Semien Mountains and the surrounding highlands is long gone. Current vegetation consists of 'a mixture of Afro-Alpine woods, heath forest, high montane vegetation, montane savanna, and montane moorland.' Species in these areas include:

> heath tree *Erica arborea*, giant lobelia *Lobelia rhynchopetalum, Solanum* sp., *Rosa abyssinica*, everlastings *Helichrysum* spp., and mosses. On ridge tops and gorge sides there are coarse grasslands with herbs *Thymus* spp., *Trifolum* spp., *Geranium arabicum*, thickets of *Rumex nervosus*, scattered *Otostegia minucci*, and creepers *Clematis simensis* and *Galium spurium*.

At one point, forests of St. John's wort *Hypericum* spp. were common at elevations of 3,000 m to 3,800 m, but few now remain.[37] Patches of remnant forests of tid (*Juniperus procera*), kosso (*Hagenia abyssinica*), and zigba (*Podocarpus gracilior*) are found at the highest altitudes and in steep, rugged landscapes unsuitable for cultivation or where cutting is hard.

The park was established in 1969, following a recommendation by a UNESCO mission. Upon its establishment, the park enclosed significant proportions of agricultural land (estimated at 22,500 hectares in size) and settlements whose populations had been there for centuries.[38] The park was set up primarily to protect three of the world's most endangered and extremely rare large mammals, the Walia ibex (*Capra walie*), the Semien or Ethiopian wolf (*Canis simensis*), and the gelada baboon (*Theropithecus gelada*). Other large mammals, such as the Bushbuck (*Tragelaphus scrptus*), klipspringer (*Oreotragus oreotragus*), hamadryads baboon (*Papio hamadryads*), black and white Colobus (*Colobus guereza kikuyuensis*), serval (*Leptailurus serval*), greater Kudu (*Tragelaphus strepsiceros)*, antelope, leopard (*Panthera pardus*), caracal (*Felis caracal*), wildcat (*Felis silvestris*), spotted hyena (*Crocuta crocuta*), and warthog (*Phacochoerus africanus*), also live in lower elevations around the park. A total of 33 species of mammals, of which 10 are endemic, have been identified in the park.[39] In addition, one of the world's endemic bird areas lies within the park.[40] Nearly 140 bird species have been recorded thus far, including 16 species endemic to Ethiopia.[41]

Other functions of the park include preserving scenic beauty and habitat diversity, restoring and rehabilitating disturbed areas, and promoting the educational, scientific, and touristic values of the park.[42] The Semien Mountains National Park was accorded international recognition when the World Heritage Convention of the UNESCO inscribed it on the World Heritage List in 1978. From its inception until 1977, expatriate wardens paid by the World Wildlife Fund and other international agencies oversaw the park. The Park was closed to the public from 1983–1999 during a 17-year civil conflict in the region. After the park had reopened, tourist visits increased from 655 in 1999 to 1,000 in 2000 to 7,000 in 2007.[43] EWCO managed the park until the responsibility was transferred to the Amhara regional state in 1997.

Until the late 1970s, over one-half of the park was used by local communities for crop cultivation and grazing. In 1979, however, the Derg military government significantly reduced the impact of human activities in the park by forcefully relocating about 1,500 farmers to lower altitudes outside the park.[44] The government stationed fifteen wildlife guards on half a dozen camps to protect the park from being accessed by the local people. The population of Walia ibex rose from less than 200 when the park was established in the late 1960s to approximately 700 to 900 in the early 1980s. From the mid-1980s to early 1990s, however, the protection of the park suffered immensely as the civil war in the north spread to the Semien region, including the park. Lack of security drove out the guards, and many park animals were killed or forced to roam outside the park. Wild animals were killed not for food or animal products but to retaliate against the harsh military government. Protection of the park remained lax following the end of the civil war; consequently, much of the park fell under cultivation and grazing uses. The numbers of Walia ibex were reduced from 400 in 1989 to 250 in 1996[45] and are now threatened due to habitat loss and hybridisation with free-ranging domestic goats.[46] Nearly 8,000 people from the surrounding communities had already settled inside the park by the end of 1995.[47] Increased settlement in and around the park has resulted in increased cultivation, the need for firewood, and hunting. Increased cultivation has led to soil degradation as people settled and cultivated on steep slopes, with some farms just below the frost limit at 3,800 m. Settlers using tree heathers and long grass for roofing have depleted resources necessary for the survival of Walia ibex. High numbers of grazing cattle and other domestic animals have had a devastating impact on the afro-alpine grassland ecosystem. With only short stubbles remaining, the soil is exposed to erosion, and the number of plant species is declining rapidly.

Road construction has also greatly affected the landscape and natural resources of the park. The alignment of the new road between Minidigebsa and Sankaber follows the 'ecologically vulnerable border zone between the plateau and the heather forests,' known as a wildlife corridor.[48] Wildlife corridors connect two or more habitats and function as an interconnected system of protection for endangered species. However, the road construction separated the Walia ibex, Klipspringer, and Gelada foraging habitat from the cliffs, decreasing their chances of survival. The roads also caused the destruction of forests and afro-alpine grassland, while earth cut from slopes or plateaus rolled down cliffs, ruining the natural beauty of the landscape and destroying rare flora.[49] In 1996, because of critical population declines of the native species, the park was added to the list of World Heritage Sites in danger.

In 1997, a new management plan for the park was initiated in conjunction with the Austrian Development Cooperation, the Ethiopian Park Management Office, and the Institute of Botany at the University of Vienna. The park management plan takes into account the International Union for the Conservation of Nature (IUCN) guidelines for national parks and the park's status as a World Heritage Site. The primary goals of the new park management plan are to conserve landscapes, ecosystems, species, and genetic variation; promote sustainable economic and human development; and provide support for research, monitoring, and education concerning conservation. Zoning management plans were designed to ensure these functions. Zoning organises the SMNP into three areas: the core area, the inner buffer, and the outer buffer zone.

The core area consists of the main wildlife habitat zone inside the park providing long-term protection to the landscape, ecosystem, and species it contains, including the endangered Walia ibex and the Klipspringer. The aim of the management plan is to provide maximum protection to endangered species and keep their habitats wholly undisturbed by banning human access to grazing, cultivation of crops, felling trees, mining, camping, and settlements. The inner buffer zone protects the core and can be used for conducting experimental research on natural resource conservation and rehabilitation. The outer zone is a development zone aimed at providing a livelihood for farming communities adjoining the park.[50]

The management plan envisions the expansion of the core area and the wildlife development zone as forests regenerate, and more land is made available by reducing human activities within the park. However, park residents have to be removed in large numbers for this to materialise. A significant proportion of the park is already under cultivation, cattle graze on large parts of the park, and the population residing within the park is increasing.[51] At present, several communities live either partially or wholly within the park. In the highlands, these communities include Mindigebsa and Adisge, Abergina, Ambaras, Argin, and Lori. In the lowlands are Adebabey, Agidamiya, Qabana, and Sera Gudela, and Angwa and Kernejan. Altogether, there are 30 villages, two of which are completely within the park's boundaries. About 5,000 households or around 28,000 people live in and around the park.[52] Consequently, there is a conflict between those who live within the park and the park management. The park management wants the park to be as undisturbed by humans as possible. The park management argues that if encroached upon, the park will become highly eroded, degraded, and devoid of wildlife like many of the other landscapes in highland Ethiopia. The park management, therefore, wishes to move villagers residing within the park. However, the park's residents claim that they have the right to use the land as their ancestors did. Villagers complain that the park management excludes them from the planning and decision-making process. It also does not let them share in the benefits of the park, such as employment opportunities and tourism. Local peoples also claim that they had been the guardians of the area before it became a national park. They believe they should reap the benefits of the park.[53] Reducing or removing human activity inside the park will therefore not be easy for the park management unless the livelihood needs and interests of the residents are incorporated in the new park management and conservation plans.

Bale Mountains National Park

The Bale Mountains National Park (BMNP) in eastern Oromiya regional state is a high-altitude plateau that is broken by many volcanic peaks, plugs, lakes, rushing rivers, and streams that descend into deep rocky gorges on their way to the surrounding lowlands below. At 2,200 square kilometres in size, the park is the largest area of afro-alpine habitat in Africa. Most of the habitat above 3,400 square kilometre is within the park. The park was established to protect Ethiopia's threatened endemic mammals in 1971, particularly the mountain nyala and the Ethiopian (Semien) wolf. Precipitation in the region is relatively copious, averaging over 1,600 mm annually. The park encloses a variety of distinctive habitats. Grasslands, woodlands, and heather moorlands dominate the northern part of the park, extending from 3,200 m to 3,600 m in altitude. From 3,700 m to 4,400 m in elevation is the central Sanetti Plateau, with afro-alpine vegetation consisting of tussock grassland and scrub covering the landscape. In the southern slope of the Sanetti Plateau, which extends from 1,500 m to 3,200 m in altitude, is the Harenna dense forest, one of the few large forests in the country that remain relatively unscathed.[54] Shallow depressions on the plateau fill with water during the wet season, forming sizeable lakes. Larger lakes, such as Garba Guracha, Hora Bachay, and Hala Weoz, contain water year-round and provide a habitat for water birds that migrate from Europe during the northern winter season. Rising from the Bale Mountains are more than forty tributaries that discharge their water in the five major rivers (the Wabi Shebelle, Weyb, Welmel, Dumal, and Genale) that drain the vast arid and semi-arid lowland areas of southeastern Ethiopia and Somalia.

The park supports a diverse range of mammals and birds. The park has 68 species of mammals of which the mountain nyala (*Tragelaphus buxtoni*), the Ethiopian wolf (*Canis semiensis*), and Menelik's bushbuck (*Tragelaphus scriptus meniliki*) are the most outstanding endemic species.[55] Two hundred sixty-five bird species have been identified in the park. Nearly three-fifths of Ethiopia's endemic birds are in the Bale Mountains – birds like Rouget's rail, blue-winged goose, spot-breasted plover, the black-headed siskin, and white-backed black tit. Several threatened bird species are also known to exist in the park, including Palled Harrier, Lesser Kestrel, Imperial Eagle, and Abyssinian Longclaw.[56] Eurasian migratory species, such as the ruddy shell duck, the golden eagle, and the chough, breed here, as does the Wattled crane. Over 900 plant species exist in the park, 17 of which are known to be endemic.[57]

One of the endemic animals in the BMNP is the Ethiopian wolf. It is among a group of unique species living in mountains that are more than 3,000 m above sea level.[58] The wolf, which was always regarded as a rare animal, was listed as needing protection in 1938. It is one of only two canid species in the *endangered* category on the IUCN Red List of Threatened Animals, the other being the North American red wolf *Canis rufus* (Table 6.2).

An estimated population of 460 to 510 adult Ethiopian wolves lived in the Bale Mountains in 1990, but their numbers dropped by 65 per cent to 75 per cent during the 1990s (80 per cent in the last ten years), and recent sightings of the wolves have also declined steadily. This decline has recently prompted the IUCN to reclassify the Ethiopian wolf as critically endangered due to its extremely high risk of extinction.

Many factors contribute to the endangerment of the Ethiopian wolf, including habitat loss and fragmentation. Increases in high altitude subsistence agriculture have lessened the amount of available habitat. The presence of large numbers of domestic dogs in Ethiopian wolf habitats is also a serious threat to wolves in the BMNP because of competition for food, disease transmission, and hybridisation. Domestic dogs are present throughout most wolf ranges in the BMNP, as well as around villages and settlements. They are kept to guard livestock against

Table 6.2 Endangered animals in Ethiopia

Critically endangered
African Wild Ass (*Equus africanus*)
Bilen Gerbil (*Gerbillus bilensis*). Endemic to Ethiopia
Black Rhinoceros (*Diceros bicornis*)
Ethiopian Wolf (Simien Jackal) (*Canis simensis*). Endemic to Ethiopia
Guramba Shrew (*Crocidura phaeura*). Endemic to Ethiopia
Harenna Shrew (*Crocidura harenna*). Endemic to Ethiopia
MacMillan's Shrew (*Crocidura macmillani*). Endemic to Ethiopia
Walia Ibex (*Capra walie*). Endemic to Ethiopia

Endangered
Gravy's Zebra (*Equus grevyi*)
Mountain Nyala (*Tragelaphus buxtoni*). Endemic to Ethiopia
Nubian Ibex (*Capra nubiana*)
Wild Dog (*Lycaon pictus*)

Vulnerable
African Elephant (*Loxodonta africana*)
Ammodile (Gerbil Family) (*Ammodillus imbellis*)
Bailey's Shrew (*Crocidura baileyi*). Endemic to Ethiopia
Bale Shrew (*Crocidura bottegoides*). Endemic to Ethiopia
Beira Antelope (*Dorcatragus megalotis*)
Cheetah (*Acinonyx jubatus*)
Dibatag (*Ammodorcas clarkei*)
Dorcas Gazelle (*Gazella dorcas*)
Glass's Shrew (*Crocidura glassi*). Endemic to Ethiopia
Large-eared Free-tailed Bat (*Otomops martiensseni*)
Lesser Horseshoe Bat (*Rhinolophus hipposideros*)
Lion (*Panthera leo*)
Moorland Shrew (*Crocidura lucina*). Endemic to Ethiopia
Morris's Bat (*Myotis morrisi*)
Mouse-tailed Bat Species (*Rhinopoma macinnesi*)
Natal Free-tailed Bat (*Mormopterusa cetabulosus*)
Nikolaus's Mouse (*Megadendromus nikolausi*). Endemic to Ethiopia
Patrizi's Trident Leaf-nosed Bat (*Asellia patrizii*)
Red-fronted Gazelle (*Gazella rufifrons*)
Rupp's Mouse (*Myomys ruppi*). Endemic to Ethiopia
Scott's Mouse-eared Bat (*Myotis scotti*). Endemic to Ethiopia
Soemmerring's Gazelle (*Gazella soemmerringii*)
Speke's Gazelle (*Gazella spekei*)
Spotted-necked Otter (*Lutra maculicollis*).
Stripe-backed Mouse (*Muriculus imberbis*). Endemic to Ethiopia

---

Source:http://iberianature.com/wildworld/guides/wildlife-of-ethiopia/endangered-animals-in-ethiopia/.

predators, mostly the spotted hyena, and are semi-feral, feeding on carrion and garbage. Genetic testing has confirmed cases of male dogs breeding with female Ethiopian wolves. The dogs often roam the wolf habitat and compete with them for rodent prey. The dogs are a vehicle for the spread of pathogens to wildlife due to their frequent travel in and out of urban areas as well as the BMNP. Rabies is a serious problem for rare carnivore populations and may be the most dangerous disease threatening the Ethiopian wolf.

When the government established the Bale Mountains National Park, it contained fewer than half a dozen settlements with a total population that did not exceed 500. By the mid-1980s, an estimated 2,500 people and their livestock permanently lived in the park, particularly in the fertile river valleys and on the vast Sanetti plateau. In the aftermath of the 1991 downfall of the Derg regime, the park was left unprotected and large numbers of people from the surrounding communities moved into the park with their animals. Even though patrolling by game guards was restored by 1993, settlements, grazing, cultivation, hunting, and logging activities inside the park continued almost unhindered. By 1994, an estimated 7,000 people had taken up permanent residency inside the park. The human intrusion resulted in the mass killing of the Mountain Nyala. The number of the Nyalas fell from a high of 2,200 in 1990 to a low of less than 600 in 1994.[59] In 2003, an estimated 40,000 people lived in settlements, including Gojera, Rira, Wege, Harenna, Ayida, Gama Taja, and Meskel-Haricho. The afro-alpine areas of Web Valley, the Central Peaks, and Sanetti are increasingly being used for grazing cattle by seasonal agropastoralists, while the Harenna forests are cleared to make room for crop cultivation.[60]

Park management has been in constant conflict with local communities ever since the creation of the park. Local communities around the park feel that the creation of the game park has denied them access to farmland, pasture, and forest resources inside the park that they consider rightfully theirs. As one resident of the region put it, 'we should not be prevented from utilizing our God-given resources to pursue our livelihood.' Clashes with park management are bound to continue so long as local communities residing in and around the park remain excluded from directly participating in the conservation and management efforts of the park. By having a say in the management of protected areas, local people obtain the right to decide what happens to the natural resources they depend on for their survival. They also acquire a vital role in society as managers and protectors of valuable resources that can sustain current and future generations.

Abijata-Shala Lakes National Park

The waters of Abijata and Shala and their surrounding areas form the Abijata-Shala National Park. This nearly 900-square kilometre park, which was established to protect the high diversity of terrestrial and aquatic birds and the scenic splendour of the areas, is home to one of the largest African colonies of the Great White Pelicans (*Pelecanus onocrotalus*). These birds, and others like White-necked Cormorants, Abdim's Storks, Sacred Ibis, Cape Wigeons, Egyptian Geese, and Speckled Pigeons use Lake Abijata as their feeding grounds and the islands on Lake Shala primarily as their breeding and roosting sites.[61] There are about 436 bird species in the park: 292 terrestrial and 144 aquatic.[62] However, the ecosystems within the park and the park, in general, are suffering from poor protection. Fields of sorghum and maize have replaced its savanna woodland acacias (*A. etbaica, A. tortilis*, and *Euphorbia abyssinica*) and bushes of *Maytenus senegalensis*. The lowering of the Bulbulla River has reduced the riverine vegetation, and overgrazing has destroyed the grassland. Much of the acacia forests around the lakes are gone. Deforestation and charcoal schemes, farms, and grazers are all making the park less habitable for wildlife.[63] In fact, the wildlife of the area has almost entirely been eliminated by habitat destruction. The area had an abundance of species in the past, including Swayne's Hartebeest, Water Bucks, Buffalo, Oryx, Giraffe, Grant's Gazelle, Greater Kudu, Klipspringer, Olive Baboons, Colobus and Vervet monkeys, Jackals, Warthog, and Spotted Hyenas. Today, there is no evidence of the presence of most of these species in the park, except the spotted hyena, warthog, and Colobus monkeys. A few ostriches live in captivity next to the main entrance to the park.

The park, which has been under development since the early 1970s, was created to protect the two lakes, the different aquatic birds, and the spectacular scenery of the area. Over 400 species of water birds are known to exist in the lakes, including several breeding colonies of Great White Pelicans, Sacred ibis, White-breasted Cormorant, Wattled Crane, Northern Shoveler, Black-winged Stilt, Little Stint, Black-necked Grebe, Greater Flamingo, Lesser Flamingo, Avocet, Ruff, Egyptian Goose, Sand Grouse, egrets, plovers, and ducks.[64] Lake Shala has diverse avifauna, especially on its islands, which serve as breeding grounds for storks, pelicans, and cormorants.[65] The park also hosts European migrant birds during the winter season in the Northern Hemisphere.

Human settlements and activities, such as fishing, hunting, grazing, farming, and removing vegetation, are not allowed within the park. In reality, however, all of these activities are taking place because the local communities refused to vacate the area they consider vital to their livelihood. The park is extensively grazed, and fields of maize and sorghum are becoming more common features of the landscape. The park has undergone severe degradation as a result of extensive tree felling for the production of charcoal. Hotsprings on the eastern shore of Lake Shala are major sources of attraction for local communities. People frequently bathe in the hot water, believing that it cures many types of ailments. Hence, the failure of the park management to take into consideration the needs and interests of the local people from the outset has rendered the protection of the park practically ineffectual. The park management has been able to tend only to a small-scale ostrich farm located at the entrance to the park. During a 2015 visit, the park had 48 guards to patrol the 900 square kilometres of 'protected' area that people living around the park consider theirs. However, with a large number of human and cattle population already inhabiting the park, effective protection of the park is going to be extremely challenging.

Nech Sar National Park

The Nech Sar National Park is a small park of 514 square kilometres located between two lakes in the southern end of the Ethiopian Rift Valley in the SNNP regional state. To the north of the park lies the ever-muddy Lake Abaya, which, at 1,160 square kilometres, is the second largest lake in Ethiopia – behind Lake Tana. To the south is Lake Chamo, a small, clear-water lake of 551 square kilometres. Enclosed by the lakes and the Amaro hills to the east (which rise to about 2,000 m) are the central plains. If seen from a distance, the plains appear to be white, giving rise to the name Nech Sar, which means white grass in Amharic. The main rains in this savanna grassland occur from March to May, with a small peak from September to November. The average annual rainfall ranges between 700 mm and 1,000 mm. The mean monthly temperatures usually vary between 25 degrees and 35 degrees centigrade.

The Nech Sar was proposed and gazetted as a national park in 1962 to protect the prolific wildlife that inhabited the area, 84 species altogether. The central plains support populations of Burchell's zebra (shown in Figure 6.2), Grant's gazelle, and the endemic Swayne's hartebeest. Other mammals commonly observed on the plains are the leopard, greater kudu, waterbuck, jackal, spotted hyena, bushbuck, Guenther's dik-dik, warthog, klipspringer, mountain reedbuck, olive baboon, Colobus monkey, and grivet monkey. Less commonly seen are lions, serval cats, wild dogs, porcupines, mongooses, civet cats, and bush pig. Large populations of hippopotamuses and crocodiles make their homes in the Abaya and Chamo lakes. Species that were found in Nech Sar but have since been eliminated include the elephant, black rhinoceros, buffalo, and oryx. Over 330 bird species are known to exist in the park, including three endemic birds: the Nech Sar nightjar, thick-billed raven, and wattled ibis. Additionally, the imperial eagle, lesser kestrel, and lesser flamingo are all listed as threatened birds of the world.[66]

This park has never been well received in the region because thousands of agropastoral communities, whose livelihoods depended on the resources of the park, were pushed off the land when the park was established. Unsurprisingly, the local communities unleashed their resentment and frustration by destroying park infrastructure, pillaging park property, killing scores of park animals, and moving back into the park by the thousands immediately following the fall of the Derg military regime in 1991. But the government later retook the park by displacing ten thousand people of the Guji cattle herding and Kore farming communities, with little compensation, from and within Nech Sar before the African Parks Foundation (APF) took over management in 2004.[67] This was done

Figure 6.2 Nech Sar National Park: Burchell's zebras standing under dry acacia trees (photo by author).

in contravention to the ILO's 1985 Convention on Indigenous and Tribal Peoples in Independent Countries, which recognises 'the rights of ownership and possession of the peoples concerned over the lands, which they traditionally occupy.'[68] At any rate, eviction of park residents did not ultimately take the pressure off the park. People continue to use park resources, but they just have to walk farther to obtain them. In fact, some have also illicitly returned to the park. Thus, the conflict between local communities and the park management continues even after the APF takeover. A growing number of cultivators and pastoral groups residing on the fringe of the park continue to graze, cultivate crops, and collect wood for fire and building. The residents of the nearby town of Arba Minch, the fast growing administrative capital of the North Omo zone, depend on the park for firewood and construction material as well.

Mago National Park

There are six wildlife conservation areas in the lower Omo River basin: two national parks, (Omo and Mago), a wildlife reserve (Tama), and three controlled hunting areas (Maze, Omo West, and Murle). Established in 1979, the Mago National Park is on the southern end of the Great Rift Valley in the Omo River Valley. It occupies an area of 2,162 square kilometres lying on the eastern bank of the Omo River. Much of the park, especially the southern portion, is a vast flat landscape that stretches out up to Chew Bahir. Savanna grass and acacia forest cover the landscape. Patches of large forest are found along rivers. The Mago and Neri rivers traverse the park and eventually join the Omo River. The park is hot throughout the year, with monthly temperatures ranging between 24 degrees and 38 degrees centigrade, while the annual precipitation range is between 400 mm and 500 mm.[69] The park is home to 81 species of mammals and more than 300 bird species, four of which – the Wattled Ibis, thick-billed raven, white cliff chat, and black-headed forest orioles – are native to Ethiopia. About 2,000 buffaloes and 200 elephants roam the park, as do lions, giraffes, leopards, Swayne's hartebeests, greater and lesser kudus, Reed bucks jackals, spotted hyenas, water bucks, Burchell's zebras, and baboons.

Omo National Park

The government established the Omo National Park in 1966 to protect wildlife similar to those found in the Mago National Park. The park is located on the western bank of the Omo River, across from the Mago National Park. It covers an area of approximately 4,068 square kilometres.[70] It is a vast open savanna 'grassland interspersed with various stands of woodland species, bush, and riverine vegetation.'[71] Seventy-three mammalian species, the largest number in any national park in the country, and 312 bird species have been identified in this park – accounting for 38 per cent of the total recorded in the country. The park is rich with 30 species of reptiles, of which crocodiles are the most common.[72] Several primate species, including the most common – olive baboon, velvet monkey, the

Abyssinian Colobus, and the patas monkey – live in the riverine forests and dense bushes along the Mui and the Omo River course. Twenty species of carnivores dwell in the park as well. The most common carnivores include the lion, spotted hyena, leopard, cheetah, and jackals. The vast open grasslands and woodlands in the park also support 26 herbivore populations, the most common of which include eland, buffalo, waterbuck, Grant's gazelle, lesser kudu, Swayne's hartebeest, greater kudu, elephant, black rhinoceros, Burchell's zebra, and giraffe.[73] The lower reaches of the Omo River were declared a UNESCO World Heritage Site in 1980, after the discovery of the earliest known fossil fragments of *Homo sapiens*, which are dated to be about 195,000 years old.[74]

Diverse cultural groups inhabit the Mago and Omo national parks – which together cover nearly one-fifth of the SNNP regional state territory. The Omo River is crucial to the survival of the various cultural groups who live along it. The groups include: the Mursi, with a population of 6,000–10,000; the Suri or Surma, with a population of around 28,000; the Nyangatom or Bume, with a population of 15,000–20,000; the Dizi, with a population of around 25,000; the Hamer and the Banna, with a population of 30,000 each; the Ari, with a population of 50–60,000; and the Me'en, with a population of around 55,000.

They all live in villages along the Omo, Mago, and Neri rivers. As discussed in Chapter 3, the economy of these semi-nomadic, agropastoral, and hunting groups is based on hunting, fishing, cattle herding, dry-season cultivation, and riverbank cultivation following the Omo and other rivers.[75] The Mursi, Surma, and Nyangatom are agropastoralists who practise flood-retreat, rain-fed agriculture, and cattle herding. These groups must migrate short distances seasonally because of the distribution of agricultural and grazing land in this semi-arid environment. The Mursi inhabit both the Omo and Mago National Park. The parks enclose about three-quarters of their resource base. The Dizi cultivate crops while the Me'en practise shifting cultivation and cattle herding.[76] These indigenous communities' livelihood has also been adversely affected by the construction of a series of dams on the Omo River, as discussed in Chapter 5.

The establishment of the parks has, thus, significantly reduced these cultural groups' access to their traditional means of livelihoods and, as a result, local communities have become resentful and resistant toward the parks.[77] Despite restrictions, local communities continue hunting wildlife in the park. Hunting has increased over the last few years, as modern weapons are more readily available. Food, leather, medicinal products, income from ivory, and defence of livestock (when killing predators) are often cited as reasons for illegal hunting. Those species most vulnerable during these times are the lesser and greater kudu, eland, buffalo, Beisa oryx, Soemmering's gazelle, Swayne's hartebeest, Gravy's zebra, elephant, cheetah, lion, spotted hyena, and jackal.[78] In 2006, the federal government contracted the management of the Omo National Park to the Dutch African Parks Foundation, otherwise known as African Parks Network (APN).[79] But unable to deny the indigenous population access to livelihood resources within the park borders, the APN gave up the park management and left Ethiopia after only two years into the management contract. In other words, the APN pulled out because its philosophy of park management devoid of human occupation became incompatible with the realities on the ground.[80] In the end, the APN also found itself flouting the ILO's Convention on Indigenous and Tribal Peoples in Independent Countries mentioned earlier in the Nech Sar National Park's case.[81]

Gambela National Park

The Gambela regional state is one of the least developed states in the country. The Baro, Gilo, Itang, Alwero, and Akobo rivers traverse the state. It is the wettest and most water-endowed region of Ethiopia. The Gambela flood plain is the largest low-lying wetland in the country.[82] It stretches from Gambela to Jikawo in the west and Gog in the south. The wet season is from May to October, with July and August registering the highest amounts of rainfall. Annual precipitation ranges from 1,000 mm in the lowlands to 1,700 mm in the highlands.[83] The Gambela National Park was proposed in 1975 to manage the middle part of the state, designated as a protected area. The park covers approximately 5,061 square kilometres or 20 per cent of the state's territory. It is between the Baro and Gilo rivers with the Baro River bordering it to the north. The park is the largest protected area in Ethiopia. Two Controlled Hunting Areas, Jikawo and Tod, are part of this protected area. Much of the topography of the area is flat with few distinguishing high grounds. Elevations range from 300 m to 400 m in the Jikawo, Akobo, Jor, and Itang districts, while elevations in much of the Abobo and Gor districts range from 400 m to 600 m. Extensive wet grassland and swamps cover the park. Unlike the rest of the districts, Godere is predominantly hilly, with elevations ranging between 600 m and 1,300 m above sea level. Gore and much of the eastern parts of Abobo and the Gambela district are covered in dense woodland and savanna vegetation. Because the Akobo, Baro, and other rivers in the region are slow moving, the low gradient of the topography often causes the banks to overflow. Hence, one finds here the most extensive swamps and marshes in the country. The perennial marshes and swamps with grasses, sedges, and scattered trees provide valuable grazing grounds for cattle herders during the dry season and refuge for wildlife. Acacia bushes and tall grasses, shemboqo (*Arundo donax*), shenkorageda (*Saccharum officinalis*), and temba (*Pennisetum petiolare*) cover the banks of the Baro and Akobo rivers. Unfortunately, recent years have seen the growth of water hyacinth in large sections of the river.[84]

The park was created primarily to protect two species of endangered wetland antelopes: the White-eared kob and the Nile lechwe. Numerous other species of wildlife also inhabit the park, including the African Buffalo, Abyssinian reedbuck, elephant, Roan antelope, lion, zebra, giraffe, bushbuck, warthog, hyena, hartebeest, olive baboon, and guereza monkey. The Gambela plains host over 300 species of birds, of which three are nearly threatened with extinction: the Shoebill, Black-winged Pratincole, and Basra Reed Warbler.[85] However, effective management of the park has yet to materialise due to lack of legal protection. At the same time, many development projects (including large-scale irrigation, dams, and roads) have been put in place that would fundamentally undermine the conservation efforts of the park.[86] The Baro-Akobo basin, which contains possibly the largest irrigable area in the country, has long been in the federal government's plan for future agricultural development. Only the lack of investment resources, infrastructure, and a market has prevented the government from implementing its plans. In 2008, BirdLife International wrote: 'Even though proposals to set up this conservation area have been planned since 1973, there has been almost no development activity. The area proposed is very large and the available infrastructure is completely inadequate to manage it effectively.'[87]

Poaching is a major problem and is seriously affecting large mammals. People poach for various economic reasons. Some groups kill wildlife for bush meat to supplement their protein requirement. Others kill for commercial products like animal skins and tusks. Civil war in southern Sudan also had a toll on wildlife. During the 1980s and 1990s, the park and the whole southwestern border areas (including Guraferda, Bebeka, Dima, Asosa, Kumurk, Begi, Tume, Gidami, and Anfilo) suffered the decimation of wildlife by automatic weapons of the Sudan People's Liberation Army (SPLA). The SPLA military activities were major reasons for the depletion of such wildlife as elephants, buffalo, eland, hartebeest, and Grant Gazelle. Firearms were readily available for local people, and large numbers of refugees moved into the park during the war. Local inhabitants customarily use bush-meat and previously hunted only with primitive weapons like spears and traps. At present, both commercial hunters and the local people use modern weapons. Wild animals are killed to protect crops and domestic animals. Animals like the Velvet monkey, Anubis baboon, porcupine, buffalo, and elephant are blamed for crop destruction, while lions and hyenas prey on domestic animals.

The woodlands and forests within the park are also being cut for construction and fuel for the growing Gambela town. There is visible diminution of park vegetation as it is frequently burnt for agriculture. The Alwero dam and the planned expansion of irrigated farms pose serious threats to the park.[88] The ongoing, and possibly increasing, land leasing to foreign investors for commercial farming will also be destructive to this ecologically rich national park in particular and the natural resources of the region in general. The government's plans to parcel out to foreign investors millions of hectares of precious, well-watered land in this state will certainly drain the swamps. The government has already resettled the native inhabitants to make way for large-scale agricultural investments.[89] The regional state wants to conserve this unique park and benefit from its economic potential, but the federal government appears to have a different idea to develop the resources of the state. Over 100,000 hectares of irrigable sites, 3.5 million hectares of land suitable for annual cropping, and nearly a million hectares of land suitable for perennial crops and grazing have been identified for development by the federal government.[90] The adverse impact of large-scale land leasing in Gambela was previously discussed in Chapter 2.

The Senkelle Swayne's Hartebeest Sanctuary

The government created the Senkelle Swayne's Hartebeest Sanctuary in 1976, specifically to protect the Swayne's hartebeest (*Alcelaphus buselaphus swaynei*), an endemic species. The sanctuary, which is only 36 square kilometres in size, is also a suitable habitat for other wildlife like the greater kudu, reedbuck, lesser kudu, as well as carnivores like the serval, spotted hyena, and common jackal. One hundred sixty-seven species of birds have been recorded.[91] The area once belonged to the Arsi Oromo, who were pushed off the land to make room for the sanctuary.

The Swayne's hartebeests were distributed widely throughout the Ethiopian Rift Valley and northwestern Somalia until the early nineteenth century. Their numbers have fallen off so low that they are now on the edge of extinction. There were close to 2,400 Swayne's hartebeests in the late 1980s. But there were only 123 at the sanctuary in the late 1990s. Only 40 or so Swayne's hartebeests remain in nearby Nech Sar National Park. During the chaotic transition from the Derg regime to the current government in 1991, guards deserted the sanctuary, and the local people destroyed the infrastructure and hunted Swayne's hartebeest, killing as many as 1,300 animals. These actions were seen as a reprisal against the sanctuary guards who had prohibited the local people from using the land.[92]

The government looked for external help to deal with the fast dwindling numbers of the Swayne's Hartebeest. In March 2002, Al Wabra Wildlife Preservation (AWWP), an environmental organisation from Saudi Arabia, agreed on a Memorandum of Understanding with the EWCO to lend emergency support to protect the few remaining animals from extinction. The group proposed to strengthen the existing ranger patrolling infrastructure to minimise incursions into the sanctuary and poaching. They also increased participation of local communities for the protection of the endangered hartebeests and provided logistical support to the EWCO in the Senkelle Sanctuary. AWWP funding employed six additional rangers from the local community for three years. Also, a security outpost enabled increased surveillance at the Palm Tree Valley in the Senkelle Sanctuary. Finally, AWWP provided a patrolling vehicle, base communication radio systems, and field equipment like binoculars for the protection of the Swayne's hartebeest. The project was completed in 2005, with long-term monitoring and heightened surveillance indicating a trend toward recovery for the Swayne's hartebeest population in Senkelle. Yearly census counts indicated that the population at Senkelle grew from less than 150 heads in 2002 to more than 220 by 2005. Moreover, poaching activities have sharply declined.[93]

Babille Elephant Sanctuary

The Babille Elephant Sanctuary (also known as the Harar Elephant Sanctuary) is in the East Hararghe Zone of the Oromiya regional state. The sanctuary covers about 7,000 square kilometres, including the valleys of the Daketa, Erer, and

Fafen well as the Gobelle and Borale rivers. They all flow southward through the sanctuary to drain into the Wabi Shebelle River.[94] The sanctuary has 202 plant species. The dominant vegetation includes acacia woodland mixed with succulents, such as *Euphorbia* spp. and *Adenia aculeate*. The tree *Erythrina burana* and the herb *Pelargonium erlangerianum* are found only in Ethiopia, and 18 other species are almost endemic to Ethiopia with records in only two or three other countries.[95] The local flora have many beneficial uses to the local people. There are fruits that can be both eaten and sold, as the 'leaves and flowers of *Ocimum basilicum* are added in "wot" (a traditional sauce) and tea or coffee, while *Ocimum lamiifolium* is used as a perfume plant.'[96] Twenty per cent (or 41 species) have medicinal properties, such as the *Sansevieria abyssinica*, which provides useful fibres. The *Vernonia galamensis*, whose seeds can be made into oil for domestic or commercial use, is also prevalent. *Acacia, Balanites aegyptiaca, Erythrina burana, Tamarindus indica*, and a few others have many uses, such as agroforestry and feed for domestic and wild animals.[97]

The Sanctuary was established in 1970 to provide protection to the rare endemic elephant sub-species *Loxdonta africana oleansie* recorded extinct in Somalia in 1928 and only currently found in this sanctuary. The sanctuary also provides a habitat for the black-maned Abyssinian lion, wild ass, greater and lesser kudu, caracal, common bush buck, genet, grivet, hamadryads baboon, leopard, salt's dik-dik, serval, warthog, various reptiles, and well over 100 bird species.[98]

Up until the mid-twentieth century, large populations of elephants and rhinos lived in various highland and lowland areas of the country. The rapid spread of firearms resulted in the disappearance of both animals from the northern and central parts of Ethiopia. Elephants were killed for their ivory, and rhinoceros were killed for their horns to make drinking cups, and for their thick skins to make shields.[99] Emperor Menelik was aware of the impacts of the destruction of the country's wildlife. He took measures to limit hunting. In the early twentieth century, he limited the wholesale killing of elephants to only one or two animals at a time. The principle was established that one of two tusks, technically the first that hit the ground, belonged to the crown. However, the ivory trade was so profitable that hunters ignored many of these decrees. This situation deteriorated during Emperor Menelik's long illness. His short-lived successor Lij Eyasu issued an order in 1913 demanding that all big-game hunting cease. Hunters were ordered to return to their native provinces or face punishment. However, due to the political instability of this period, the order was largely ineffective.[100] Since then, the number of elephants has fallen drastically, their range has shrunk, and the remaining habitat is being destroyed. Extensive encroachments of human settlements, livestock, and biofuels production have increasingly become a severe challenge for the future existence of elephants in the Babille Elephant Sanctuary.[101] The size of the sanctuary has decreased by two-thirds and elephant numbers have declined from 300 or more in the 1970s to less than 150 at present. This drop in numbers is not only due to loss of their habitat but also due to poaching. Farmers also kill elephants for destroying their crops, which take many months and much labour to produce.

Yabelo Game Sanctuary

The Yabelo Game Sanctuary is a 250,000-square kilometre wildlife reserve located in southern Ethiopia, to the east of the Yabelo town in the Borana Zone of the Oromiya regional state. Acacia savanna, including *A. drepanolobium, A. brevispica,* and *A. horrida,* dominates the slopes, while patches of *Balanites aegyptiaca, Commiphora,* and *Terminalia* cover areas at the lower elevations. The indigenous *Juniperus procera* and *Olea europea cuspidata* forests covered higher elevations in the past, but no longer exist today. The government established the sanctuary primarily to protect the Swayne's hartebeest. This sanctuary, however, has been almost completely overrun by livestock development and the hartebeest is almost extinct in the area. Illegal hunting has also wiped out the spotted cats and ostriches. There is now little other wildlife there, except for some important endemic bird species. The Yabelo area provides habitat for several endemic bird species in Ethiopia, including the globally threatened Abyssinian Bush-crow, White-tailed Swallow, Digodi Lark, Vulturine Guinea fowl, White-crowned Starling, D'Arnaud's Barbet, Acacia Tit, Pygmy Falcon, Yellow-necked Spurfowl, Francolin, Crested Francolin, Black-billed Wood-hoopoe, Tropical and Slate-colored Boubou, Bare-Faced Go-Away-Bird, Bare-eyed Thrush, Purple Grenadier, and Black-capped and Grey-capped Social-weavers. Numerous birds of prey make their home around the Yabelo sanctuary, including Lappet-faced and White-headed vultures, and Wahlberg's eagles.[102]

Conclusion

Most of Ethiopia's protected areas face a serious threat from human-induced activities. There is a trend of overexploitation of wildlife, to the point that species are being displaced, disappearing altogether, or being constricted to small and marginal habitats, or within conservation areas. The loss of these species leads to disruption and simplification of the ecosystems adapted to local conditions, thus disrupting ecological processes. The inability of one species to survive and replicate leads to a downward spiral of the entire ecosystem. This is especially a problem among larger species.[103]

The Ethiopian Wildlife Conservation Authority (EWCA) – formerly Organization – manages the conservation of protected areas. Established in 1965, it has overseen the creation of all the country's national parks, wildlife sanctuaries, wildlife reserves, and controlled hunting areas. The government created all of these protected areas with little or no consideration of their impact on the livelihoods of the people living in and around them. They were established without the participation or consent of local communities and often involved their forced eviction or displacement. The EWCA adopted an exclusionary protected-area policy that mandated that 'all kinds of human use of that area, such as settlement, exploitation of natural resources, grazing of livestock, and mining are prohibited except as required for the management of the wildlife and conservation.'[104] The adoption of this strict conservation policy has been controversial as previous communally owned and accessed natural resources are brought under government control and management. As a result, indigenous peoples are involuntarily pushed out or denied access to resources from the land they consider their ancestral domain. In most instances, conservation rules are largely ignored or resisted. Periodic use of force to remove those individuals or communities encroaching on the national parks has only served to infuriate people and to alienate further legitimate residents. The exclusion of people who depend on harvesting resources from protected areas has, thus far, been detrimental to the material and spiritual welfare of the people, as well as to the sustainability of the parks and their biodiversity. Consequently, past conservation programmes were unable to realise significant measures of success in protecting national parks and wildlife sanctuaries.

As Higgins-Zogib and others put it, it is the 'approach towards protected areas establishment, management, and governance that has been at the root of the problem.'[105] The command and control method of managing wildlife using the gun to fight poachers and encroachers is not going to protect national parks. Conservation of protected-area resources in Ethiopia ought to address the needs of both humans and wildlife. The IUCN defines conservation as 'the management of human use of organisms or ecosystems to ensure each use is sustainable. Besides sustainable use, conservation includes protection, maintenance, rehabilitation, restoration, and enhancement of populations and ecosystems.'[106] This definition encompasses 'both the utilization and sustained yield' of the natural resources of protected areas in regard to humans. Conservation of protected areas in Ethiopia must thus make allowances for the utilisation of natural resources while at the same time preserving them 'in as natural a state as possible.'[107]

In other words, there ought to be 'a balance between [conservation] and economics that recognizes both the desire for inviolable natural enclaves and the needs of the human population.'[108] Unless protected areas provide economic and cultural services to the people around them, they are bound to disappear if for no other reason than solely because of the pressure of rapidly growing population. Therefore, legislation enabling the participation of local communities in the development and management of protected areas would be a huge step forward, as it would not only allow for better understanding of the residents' needs but would also involve them in the solutions while providing them with tangible benefits.[109] As Agarwal and Narain have underscored, based on the Indian experience, 'People will care for their environment only if they have the legal right to manage it and to its products.'[110] In this regard, EWCA's new role ought to be providing material and technical support to local/indigenous communities so that they may establish their protected areas, and may design and implement sustainable land and resource use policies.

Notes

1 Gille A. Huxley, L. Swift Monod and E. B. Worthington. 1963. *The Conservation of Nature Natural Resources in Ethiopia*, Paris: UNESCO/NS/NR/47, p. 13.
2 Richard Pankhurst. 1992. The history of deforestation and afforestation in Ethiopia prior to World War II. *Ethiopian Institute of Development Research* 2, 2: 59–77.
3 Luisa Maffi and Ellen Woodley. 2010. *Bio-cultural Diversity Conservation: A Global Sourcebook*, London: Earthscan, pp. 6–7.

4 Tewolde Berhan Gebre-Egziabher. 1991. Diversity of Ethiopian flora. In: *Plant Genetic Resources of Ethiopia*, J. M. M. Englles, J. G. Hawkes, and Melaku Worede (eds.), Cambridge: Cambridge University Press, pp. 75–81.

5 Jonathan Mckee. 2007. *Ethiopia: Country Environmental Profile*, p. 40.

6 J. C. Hillman. 1993. *Ethiopia: Compendium of Wildlife Conservation Information*. The Wildlife Conservation Society International, New York Zoological Society, New York, Ethiopian Wildlife Conservation Organization, Addis Ababa, p. 19; Ministry of Water Resources (MoWR) and National Meteorological Services Agency (NMSA). 2001. *Initial National Communication of Ethiopia to the United Nations Framework Convention on Climate Change (UNFCCC)*, June, Addis Ababa, Ethiopia, p. 36.

7 Ethiopian Wildlife and Natural History Society (EWNHS). 1996. *Important Bird Areas of Ethiopia: A First Inventory*, Addis Ababa: EWNHS, p. 35; United Nations Educational, Scientific, and Cultural Organization (UNESCO). 2004. *National Water Development Report for Ethiopia*. Final Report, Addis Ababa, MoWR, December, UN-WATER/WWAP/2006/7. Available online at http://unesdoc.unesco.org/images/0014/001459/145926e.pdf (accessed 26 November 2010).

8 Tesfaye Hundessa. 1996. Utilization of Wildlife in Ethiopia. *Walia* 17, pp. 3–9.

9 Hillman, p. 12.

10 Huxley et al., p. 18.

11 Jane Graham. *National Parks in Ethiopia's Southern Region*, Awassa, Agricultural Bureau, Southern, Nations, Nationalities and Peoples Region, no publication date.

12 Tesfaye Hundessa, pp. 3–9.

13 UNESCO World Heritage Committee and the Ethiopian Wildlife Conservation Organization (EWCO). 1986. *Management Plan: Semien Mountains National Park and Surrounding Rural Areas*, Addis Ababa: Ministry of Agriculture, p. 10; MoWR. 1997.*Abbay River Basin Integrated Development Project*. Volume V: Environment, Addis Ababa, MoWR, p. 5.4

14 Biodiversity and Protected Areas – Ethiopia, *Earth Trends 2003*.

15 *Earth Trends 2003*.

16 Central Statistical Agency (CSA). 2011. *Statistical Abstract 2011/2012*, Addis Ababa: CSA, p. 7. Available online at http://ewnhs.org.et/wp-content/uploads/downloads/2011/03/Biodiversity-Hotspots-of-Ethiopia.pdf (accessed 4 November 2014).

17 Ministry of Natural Resources Development and Environmental Protection (MNRDEP). 1993. *Ethiopian Forestry Action Program*. Final Report, Volume II: The Challenge for Development, Addis Ababa, MNRDEP, p. 82; Shibru Tedla. 1995. Protected Areas Management Crisis in Ethiopia. *Walia*16, pp. 17–30; p. 20.

18 EWNHS, p. 49.

19 Ethiopian Valleys Development Studies Authority (EVDSA). 1989. *Master Plan for the Development of Surface Water Resources in the Awash Basin*. Final Report, volume 9, Addis Ababa, p. 9.

20 EWNHS, p. 137.

21 Ibid., p. 9.

22 Paulos Dubale. 2001. Soil and water resources and degradation factors affecting productivity in Ethiopian highland agro-ecosystems. *Northeast African Studies* 8, 1: 27–51; p. 30.

23 Michael J. Jacobs and Catherine A. Schloeder. 2001. *Impact of Conflict on Biodiversity and Protected Areas in Ethiopia*, Washington, DC: World Wildlife Fund, pp. 19–20.

24 Ibid., 25.

25 Ibid., p. 27.

26 Krishna B. Ghimire, and Michael P. Pimbert. 1997. Social change and conservation: an overview of issues and concepts. In: *Social Change and Conservation: Environmental Politics and Impacts of National Parks and Protected Areas*, Krishna B. Ghimire and Michael P. Pimbert (eds.), London: Earthscan, pp. 1–45; p. 21.

27 Jacobs and Schloeder, pp. 19–20.

28 Camerapix Publishers International. 1995. *African Wildlife Safaris,* Nairobi, Kenya: Camerapix, p. 134.

29 Shibru, p. 22.

30 Ann Gibbons. 2009. *Ancient Skeleton May Rewrite Earliest Chapter of Human Evolution, Science Now,* 1 October. Available online at www.sciencemag.org/news/2009/10/ancient-skeleton-may-rewrite-earliest-chapter-human-evolution (accessed 10 July 2016).

31 P. D. Moehlman, H. Yohannes, R. Teclai, and Fanuel Kebede. 2008. Equus africanus. In: *IUCN Red List of Threatened Species.* Available online at http://en.wikipedia.org/wiki/African Wild Ass (accessed 14 March 2011).

32 Fanuel Kebede. 1999a. Equus africanus. In: *IUCN Red List of Threatened Species.* Available online at www.iucnredlist.org/apps/redlist/details/7949/0 (accessed 14 March 2010).

33 Fanuel Kebede. 1999b. *Ecology and Conservation of the African Wild Ass (Equus africanus) in the Danakil, Ethiopia.* MSc thesis, Durrell Institute of Conservation and Biology, University of Kent at Canterbury, UK, p. 7; Fanuel Kebede. 1999a.

34 Fanuel Kebede, L. Berhanu, and P. D. Moehlman. 2007. *Distribution and Population Status of the African Wild Ass (Equus africanus) in Ethiopia.* Report to Saint Louis Zoo.

35 Fanuel Kebede. 1999a. Yichilal Fenta. 1995. *Wildlife Resources in Ethiopia with Particular Reference to Mammals.* Geography senior thesis, Addis Ababa University, p. 17.

36 Fanuel Kebede. 1999b, p. 7.

37 Javiar Beltran. 2000. Case study 6: Semien Mountain National Park, Ethiopia. In: *Indigenous and Traditional Peoples and Protected Areas: Principles, Guidelines, and Case* Studies, World Commission on Protected Areas (WCPA), p. 79.

38 United Nations Environmental Program (UNEP). 2009. *Semien Mountains National Park,* UNEP: New York, 2009.

39 Shibru Tedla, p. 21.

40 A. J. Stattersfield, M. J. Crosby, A. J. Long, and D. C. Wege. 1998. *Endemic Bird Areas of the World: Priorities for Conservation,* BirdLife International, Cambridge, UK.

41 J. Hillman. 1993. *Ethiopia: Compendium of Wildlife Conservation Information.* Vol. 1. NYZS / WCSI, New York Zoological Park, Bronx, NY, p. 10.

42 EWCO. 1991. *Semien Mountains National Park,* Situation Report to UNESCO, Ethiopia.

43 UNEP. 2009. *Semien Mountains National Park,* New York: UNEP.

44 UNESCO and EWCO. 1986. *Management Plan: Semien Mountains National Park and Surrounding Rural Areas,* Addis Ababa, Ministry of Agriculture, p. 56.

45 D. M. Shackleton. 1997. *Wild Sheep and the Relatives.* IUCN/SSC Caprinae Specialist Group, IUCN, Gland, Switzerland and Cambridge, UK.

46 B. Nievergelt, T. Good and R. Guttinger. 1998. A survey of the Flora and Fauna of the Semien Mountains National Park, Ethiopia, Special Issue of *Walia,* Addis Ababa, Ethiopia.

47 EWNHS, p. 100.

48 Marco Keiner. Towards a New Management Plan for the Semien Mountain National Park. Available online at http://e-collection.library.ethz.ch/eserv/eth:26720/eth-26720-01.pdf (accessed 10 July 2016).

49 Ibid.

50 Hans Hurni and Eva Ludi. 1994. *Reconciling Conservation with Sustainable Development,* Center for Development and Environment, University of Berne, Switzerland. Available online at www.cde.unibe.ch/Tools/pdf/afr22_part1.pdf (accessed ------); Marco Keiner, Towards a New Management Plan for the Semien Mountain National Park. Available online at http://e-collection.ethbib.ethz.ch/ecol-pool/bericht/bericht_300.pdf (accessed 10 July 2016).

51 Marco Keiner. Towards a New Management Plan for the Semien Mountain National Park. Available online at http://e-collection.ethbib.ethz.ch/ecol-pool/bericht/bericht_300.pdf (accessed 10 July 2016).

52 Beltran, p. 80.

53 Ibid., p. 83.

54 L. A. Laverenchenko, A. N. Milishnikov, V. M. Aniskin, A. A. Warshavsky, and Woldegabriel Gebrekidan. 1997. The genetic diversity of small mammals of the Bale Mountains, Ethiopia. *SINET: Ethiopian Journal of Science* 20, 2, p. 213.

55 Ethiopian Mapping Authority. 1988. *National Atlas of Ethiopia*, Addis Ababa, Ethiopian Mapping Authority, p. 60). Fourteen of the birds and eleven of the mammals endemic to the country are found in this park. IUCN. 1990. *Biodiversity in Sub-Saharan Africa and Its Islands: Conservation, Management, and Sustainable Use*, Occasional Papers of the IUCN Species Survival Commission, IUCN – The World Conservation Union, p. 32.

56 A Glimpse of Biodiversity Hotspots in Ethiopia. Available online at http://ewnhs.org. et/wp-content/uploads/downloads/2011/03/Biodiversity-Hotspots-of-Ethiopia.pdf (accessed 4 November 2014).

57 Woldegebriel Gebre Kidan. 1996. The Status of Mountain Nyala (*Trigelaphus buxtoni*) in the Bale Mountains National Park, 1986–1994. *Walia* 17, p. 27.

58 The source of information on the Ethiopian Wolf is: IUCN. 1997. *The Ethiopian Wolf: Status Survey and Conservation Action Plan*, IUCN/SSC (Canid Special Group), Compiled and edited by Claudio Sillero-Zubiri and David Macdonald.

59 Woldegebriel Gebre Kidan, pp. 35–6.

60 Ethiopian Wildlife Conservation Authority (EWCA). 2007. *Bale Mountains National Park: General Management Plan 2007, 2017*, Addis Ababa: EWCA, p. 61.

61 Michele L. Thieme, Robin Abell, Melanie L. J. Stiassny, and Paul Skelton. 2005. *Freshwater Ecoregions in Africa and Madagascar: A Conservation Assessment*, Washington, D.C.: World Wildlife Fund, p. 183; Emil K. Urban. 1969. Ecology of water birds of four Rift Valley lakes in Ethiopia. *Ostrich* 40 (Suppl. 1): 315–22; p. 320. Available online at www.tandfonline.com/doi/abs/10.1080/00306525.1969.963 9131 (accessed 10 July 2016).

62 Lemlem Sissay. 2003. Biodiversity potentials and threats to the southern Rift Valley lakes of Ethiopia. In: *Wetland of Ethiopia: Proceeding of a Seminar on the Resources and Status of Ethiopia's Wetlands*, Yilma Dellelegn and K. Geheb (eds.), IUCN, Gland, p. 116.

63 EVDSA. 1992a, p. 39.

64 EWNHS, p. 125.

65 Herco Jansen, Huib Hensdijk, Dagnachew Legesse, Tenalem Ayenew, Petra Hellegers, and Peter Spliethoff. 2007. *Land and Water Resources Assessment in the Ethiopian Central Rift Valley*, Alterra-rapport 1587, Alterra, Wageningen, The Netherlands, p. 23.

66 Agricultural Bureau, Southern Nations, Nationalities and Peoples Region. *National Parks of Ethiopia's Southern Region*, Awassa, Agricultural Bureau, no date, pp. 7–8.

67 Larry Thompson. 2005. Ethiopia: local people burned out of homes to make way for national park, *Refugees International*, Washington, DC, April. Available online at www .refworld.org/docid/47a6eebb0.html (accessed 10 July 2016).

68 Native Solutions to Conservation Refugees. 2011. *The Take-Over of Ethiopia's Omo National Park by African Parks Foundation of the Netherlands*, pp. 1–4. Available online at www.danadeclaration.org/pdf/omotakeover.pdf (accessed 10 July 2016).

69 EWNHS, p. 221.

70 Matt Philips and Jean-Bernard Carillet. 2006. *Ethiopia and Eritrea*, Third Edition, Guilford, CT: Lonely Planet, p. 211.

71 EWNHS, p. 223.

72 Zelalem Tefra Ashenafi. 1994. *Management Plan of Omo National Park*, Addis Ababa: Ethiopian Wildlife Conservation Organization, pp. 36–7.

73 Ibid., pp. 34–5.
74 Matt Philips, Jean-Bernard Carillet, and Zelalem Tefra Ashenafi. 1994. *Management Plan of Omo National Park*, Addis Ababa: Ethiopian Wildlife Conservation Organization, p. 11.
75 Adam W. M. Green. 1990. *Development: Environment and Sustainability in the Third World*, New York: Routledge, p. 186.
76 Native Solutions to Conservation Refugees, pp. 1–4.
77 D. Turnton. 1987. The Mursi and National Park development in the lower Omo Valley. In: *Conservation in Africa: People, Policies and Practices*, D. M. Anderson and R. H. Grove (eds.), Cambridge: Cambridge University Press, pp. 169–86.
78 Jacobs and Schloeder, pp. 19–20.
79 African Parks Foundation was 'founded by Paul van Vlissingen, the Chairman of the global retail giant Makro Retail. African Parks Foundation manages parks in Zambia, Malawi, South Africa, Democratic Republic of Congo, Sudan, and Ethiopia and is reportedly looking at managing more. The revenue from these parks accrues to their projects, and is put towards opening more parks.' Available online at http://en.wikipedia. org/wiki/Omo_National_Park#cite_note-4 (accessed 10 July 2016).
80 Matthijs Blonk. 2011. Indigenous peoples versus the 'business model': why African Parks Networks is pulling out of Ethiopia, *Ecologie and Ontwikking*, a magazine of IUCN-NL, September 2008. Available online at www.matthijsblonk.nl/paginas/ AfricanParksEthiopieEng.htm (accessed 19 March 2011).
81 International Labor organization (ILO). 2006. *Convention Concerning Indigenous and Tribal Peoples in Independent Countries*, Geneva: ILO. Available online at www.ilo. org/ilolex/cgi-lex/convde.pl?C169/ (accessed 19 March 2011).
82 EWNHS, p. 108.
83 Ibid., p. 109.
84 Ibid., p. 111.
85 United Nations Educational, Scientific, and Cultural Organization, World Water Assessment Program. 2004. *National Water Development Report for Ethiopia*. Final Report, Addis Ababa, Ministry of Water Resources, December 2004, UN-WATER/ WWAP/2006/7, EWNHS, p. 111. Available online at http://unesdoc.unesco.org/ images/0014/001459/145926e.pdf, (accessed 26 November 2010).
86 EWNHS, p. 112.
87 BirdLife International. 2008. *BirdLife's Online World Bird Database: The Site for Bird Conservation*. Version 2.1. Cambridge, UK: BirdLife International. Available online at www.birdlife.org (accessed 16 August 2010).
88 Ibid.
89 Fred Pearce. 2012. *The Land Grabbers: The New Fight Over Who Owns the Earth*, Boston: Beacon Press, pp. 5–6.
90 Water Resources Development Authority, Transitional Government of Ethiopia. 1995. *Survey and Analysis of the Upper Baro-Akobo Basin*. Final Report, Volume I – Main Report, Addis Ababa: Addis Resources Development-GEOSERV (ARDCO-GEOSERV), pp. 24 and 31.
91 M. Mattravers and M. Netsereab 1994. *The Senkelle Swayne's Hartebeest Sanctuary Management Plan*, EWCO, Mimeo, Addis Ababa; Cited in Nobuko Nishizaki. 2004. Resisting Imposed Wildlife Conservation: Arsi Oromo and the Senkelle Swayne's Hartebeest Sanctuary. *African Studies Monograph* 25, 2: 61–77.
92 Nishizaki, pp. 61–77.
93 Al Wabra Wildlife Preservation (AWWP) is an organisation established by a wealthy Saudi Arabian – Saoud Bin Mohammed Bin Ali Al Thani – to preserve rare and endangered animals. Available online at http://awwp.alwabra.com/index.php/content/ view/163/41/ (accessed 10 July 2016).
94 BirdLife International. 2011. *Important Bird Areas factsheet: Babille Elephant Sanctuary*. Available online at www.birdlife.org (accessed 18 March 2011).

95 Demel Teketay. 1995. Floristic composition of Dakata Valley, Southeast Ethiopia: an implication for the conservation of biodiversity. *Mountain Research and Development* 15, 2, p. 185.

96 Ibid.

97 Ibid.

98 Ibid.; Philip Briggs. 2009. *Ethiopia: The Bradt Travel Guide*, Guilford, CT: The Globe Pequot Press, pp. 426–7.

99 Augustus B. Wylde. 1901. *Modern Abyssinia*, London: Methuen, p. 449.

100 Richard Pankhurst, *Ethiopian Environmental History: Lions and Elephants in Ethiopian History II*, Addis Ababa, p. 7.

101 Yirmed Demeke and Negusu Aklilu. 2008. Alarm bell for biofuel development in Ethiopia: the case of Babille Elephant Sanctuary. In: *Agrofuel Development in Ethiopia: Rhetoric, Reality and Recommendations*, T. Heckett and Negusu Aklilu (eds.), Addis Ababa: Forum for Environment in partnership with Horn of Africa Regional Environmental Center/Network, pp. 83–113.

102 EWNHS, p. 186; Briggs, pp. 502–3; BirdLife International. 2011. *Important Bird Areas Fact Sheet: Yabelo Sanctuary*, Ethiopia. Available online at www.birdlife.org/datazone/sitefactsheet.php?id=6298 (accessed 17 March 2011).

103 J. C. Hillman. 1983. *Ethiopia Compendium of Wildlife Conservation Information*, Vol. 1, Addis Ababa: Ethiopian Wildlife Conservation Organization, p. 9.

104 Jacobs and Schloeder, p. 15.

105 Liza Higgins-Zogib, Nigel Dudley, Stephanie Mansourian, and Surin Suksuwan. 2010. Safety net: protected areas contributing to human well-being. In: *Arguments for Protected Areas: Multiple Benefits for Conservation and Use*, Sue Stolton and Nigel Dudley (eds.), London: Earthscan, p. 128.

106 Hillman. 1983. p. 10.

107 Ibid.

108 Jonathan S. Adams and Thomas O. McShane. 1996. *The Myth of Wild Africa: Conservation without Illusion*, Berkeley: University of California Press, p. 176.

109 Jacobs and Schloeder, p. 13.

110 Anil Agarwal and Sunita Narain. 1989. Towards green villages: a strategy for environmentally sound and participatory development. Quoted in Richard Douthwaite. 1999. *The Growth Illusion: How Economic Growth Has Enriched the Few, Impoverished the Many and Endangered the Planet*, Gabriola Island, Canada, p. 259.

7 Conclusion

Human economies depend on the extraction of natural and environmental resources: soil, water, pasture, forests, wildlife, and minerals. The well-being of a given society depends on the quality and quantity of its natural and environmental resources and how it manages them in the pursuit of achieving sustained economic development and improvements in human welfare. Unsustainable or improper human activities reduce both the quality and quantity of available natural and environmental resources, such as soil, water, forest, pasture, and biodiversity. There is much to worry about Ethiopia's environment and resource base.

At present, Ethiopia's population is growing at close to 3 per cent a year. While the pace is expected to slow down in the future, growth will remain substantial because about 43 per cent of its population is under 15 years of age. Rural people still want to have more children. The reason is simple; children are the basis of livelihood. In an overwhelmingly agrarian Ethiopia, every child is born with a pair of hands and contributes to the livelihood of the family. Children participate in cattle herding, fetching water, collecting firewood, weeding, seeding, and harvesting. Also, rural families have no social security at old age; they count on their children to take care of them. As a result, Ethiopia's population is expected to be more than double from about 96 million at present to 180 million or more by the middle of this century. A growing population contributes to increasing levels of land conversion, resources extraction, and deterioration in the quality and sustainability of the environment. The future of the country depends on balancing the material needs of its citizens with the sustainable use of its natural resources.

Despite concerted efforts on the part of the government and rural communities around the country, a vast amount of topsoil is washed out of farm fields, biodiversity is declining, forests are being destroyed at a rapid rate, and lake ecosystems are in peril. Unless assiduously tackled, these problems could grow considerably worse and will significantly diminish the benefits that future generations of Ethiopians will obtain from the country's natural capital and ecological services. The challenge for the country is how to provide its citizens with adequate food, shelter, education, health care, and other essential resources for a secure, healthy life while finding ways to live sustainably over the long run without diminishing its natural resources and critical ecological services on which all life hinges.

Land is the most important natural resource in Ethiopia; it is the basis for agricultural production, which feeds and sustains the vast majority of the country's population. The sector is expected to contribute to feeding the growing population, to foster overall economic development, and to provide employment. The challenges surrounding agriculture in Ethiopia are numerous and varied. There are environmental factors that all Ethiopian farmers face: drought, floods, pests, diseases, and the like. Unsustainable land use in rural areas is causing land degradation, including soil erosion, nutrient depletion, salinity, and loss of biodiversity. The most efficient way to prevent or decelerate soil degradation is to keeping soil covered with appropriate vegetation and maintain adequate cover of organic matter in the topsoil. Agroforestry is especially effective in reducing erosion and improving soil quality. Trees may compete with crops for land but if properly selected, they can benefit the soil by enhancing the fertility and quality of the soil. Other measures of minimising soil erosion, such as terracing and contour ploughing, have to be practised to reduce the worst erosion effect from rain runoff. The country has to effectively handle the problem of declining soil fertility and the constraints associated with climate variability by investing in appropriate irrigation, soil conservation, and rehabilitation programmes. The country should also invest in human resources development and rural industry to provide alternative employment to reduce the pressure on farmland and other natural resources.

Pastoralism is an extensive form of land use in Ethiopia, and approximately half of the country is estimated to be suitable only for animal husbandry. Pastoralism is the principal source of livelihood for the vast majority of the people living in semi-arid and arid regions of the country. Pastoral ecosystems in the country are deteriorating, and pastoral ways of life are under severe threat. Often, pastoralists are blamed for degrading their rangelands by taking too much out of them. However, as the cases in Chapter 3 showed, the introduction of new forms of land use and the displacements of pastoral communities into more marginal lands are the principal causes of rangeland degradation in the country. The government is pushing pastoralists to settle down on a permanent basis by promising them to provide improved livestock breeds, improve the provision of mobile and marketing services, develop water resources for livestock and household consumption by diverting rivers and drilling boreholes, improve pastureland, and develop irrigation schemes.[1] But many question the effectiveness of the government's resettlement policy because it goes against pastoral strategic mobility, which is the basis for efficient use and management of pastoral scarce resources.

Forests are critical in maintaining ecological balance, sustaining biodiversity, moisture regime, reducing soil erosion, and absorbing gases. Unfortunately, considerable damages have already been done to the country's forest resources, especially to those found in the northern highland regions of the country. Many of the remaining forests in these regions are so fragmented that they no longer sustain the diversity of life previously associated with them. Deforestation has been and continues to be caused by a variety of human-induced activities: farming, energy production, construction, mining, and settlements. Federal and regional laws to protect forest resources abound but such laws are far from being effective. There is a need for both the federal and regional governments to work with local communities to prevent the further destruction of the country's forests and woodlands. It is of utmost importance that any programme designed to promote sustainable forest management and conservation place local people at the centre of the scheme. Local people themselves must make decisions that affect local livelihoods, not by the political powers above. There may also be a need for local community-supported forest zoning regulations that distinguish between protected forests and production forests. A production forest is where timber and other products can be harvested on a sustainable basis while a protected forest should not be exploited. Sustainable forest management requires a balance between harvesting wood and allowing time for regeneration. A protected forest will continue providing ecosystem services that underpin both life and livelihoods.

Ethiopia has considerable freshwater resources and a high potential for hydropower development. Its highlands are the water towers of the Horn of Africa, the sources of the White Nile, the Blue Nile, the Juba River, and the Turkana Lake that quench the thirst of a combined population of nearly 200 million in Sudan Egypt, Somalia, and Kenya. But the country is faced with many water-related challenges: water shortages in some regions, frequent floods, erosion and sedimentation, salinity, the spread of water-borne diseases related to water embankments and irrigation, declining fisheries catch due to over-exploitation, and environmental and water quality problems. The shores of the country's lakes are covered by algae and phytoplankton, a sign of water that is eutrophic or overloaded with nutrients. Algae and phytoplankton bloom, die, and decompose, using up oxygen, resulting in the suffocation of aquatic lives in the lake. The continuing destruction of watersheds and wetlands are exacerbating water scarcity in some regions. Watersheds in the country are increasingly populated and cultivated, resulting in reduced water infiltration and storage and increased soil erosion.

Clearly, continuing population growth, expanding irrigation schemes, and increasing urbanisation and industrialisation are bound to put heavy pressure on the country's water resources. Household water use is expected to increase as the country's population continues to grow, and access to water increases in demand. Industrial sector water demands will also increase as the sector continues to grow and consume increasing amounts of water for processing, cooling, boiler feed, and washing. Hence, the need to pay attention to conservation and protection of water resources is crucial. The country's water resources need to be well protected to minimise the growing pressure on water by adopting sustainable water management practices. Such management practices may include: minimising system losses and improving efficiencies; recharging ground water aquifers by reforesting watersheds; treating waste water to reduce water pollution; preserving wetlands;

building catchment dams to capture rainwater; reducing nutrient and effluent discharges into streams, lakes, canals, and reservoirs; introducing efficient irrigation methods; recycling and reusing wastewaters; and pricing water to cover the costs of production and maintenance.

Water is a geopolitical resource in the Horn of Africa. The absence of river-basin-wide cooperation in sharing water resources is generating tension between riparian countries in the region. A basin-wide cooperation is clearly needed to prevent potential conflicts as well as to promote trans-boundary water resources development and management. Such cooperation would allow the establishment of governing principles for trans-boundary water-sharing and institution-building for data-gathering and monitoring networks; the protection and rehabilitation of watersheds; and the formation of tribunals to settle disputes.

Ethiopia is home to diverse wildlife and avifauna species. Unfortunately, the country's rich biological diversity has been and continues to be diminished by detrimental human activities. Top-down approaches to conservation, where the federal or regional government owns and controls protected areas, have proven to be inadequate to safeguard wildlife. State management of parks and other protected areas has alienated local communities from livelihood resources. Unless protected areas provide economic and cultural services to the people living in and around them, they are bound to continue deteriorating. The most effective means to protect national parks and wildlife reserves is to grant exclusive territorial rights to local communities that depend on these ecosystems for their livelihood. Local communities are knowledgeable and better situated to manage their environment and natural resources. Therefore, federal and regional government institutions should recognise and respect such knowledge and involve local people in decision-making processes that affect the environment and natural resources.

The relationship between society and environment is inherently interwoven. A degraded environment can undermine the means of livelihood. The continued degradation of the environment and natural resources of the country will inevitably result in the diminution of the quality of life for future generations. The future of Ethiopia's livelihood security will depend on how its soil resources, rangelands, forests, and water are managed at present and in the future. The country cannot afford to be nonchalant as its natural systems are depleted and degraded. It must strive to utilise its renewable resources in a way such that, in the words of Herman Daly, 'harvesting rates do not exceed regeneration rates and waste emissions do not exceed the renewable assimilative capacity of the local environment.'[2] There are also lessons to benefit from the past. Jared Diamond's work *Collapse* tells the story of great societies throughout human history that collapsed because they overextended their natural resource base.[3] The country ought to make concerted efforts to develop a sound conservation system that not only safeguards its natural resources but also serves as a basis for climate change mitigation and adaptation.

Notes

1 Federal Democratic Republic of Ethiopia. 2010. *Growth and Transformation Plan, 2010/11–2014/15*, Addis Ababa.
2 Herman E. Daly. 1996. Sustainable growth? No thank you. In: *The Case against the Global Economy and For a Turn Toward the Local*, Jerry Mander and Edward Goldsmith (eds.), San Francisco: Sierra Club Book, p. 196.
3 Jared Diamond. 2011. *Collapse: How Societies Choose to Fail or Succeed*, New York: Penguin Books.

Bibliography

Abate, Teferi. 1995. Land redistribution and inter-household relations: the case of two communities in northern Ethiopia. *Ethiopian Journal of Development Research* 17:23–40.

Abebe, Mesfin. 1998. *Nature and Management of Ethiopian Soils*, Haramaya: Haramaya University Press.

Abule, E., H. A. Snyman, and G. N. Smit. 2005. Comparisons of pastoralists' perceptions about rangeland resource utilization in the Middle Awash Valley of Ethiopia. *Journal of Environmental Management* 75: 21–35.

Adams, Jonathan S., and Thomas O. McShane. 1996. *The Myth of Wild Africa: Conservation without Illusion*, Berkeley: University of California Press, p. 176.

Adams, W. M. 1990. *Green Development: Environment and Sustainability in the Third World*, New York: Routledge.

Adelson, Glenn, James Engell, Brent Ranalli, and K. P. Van Anglen (eds.). 2008. *Environment: An interdisciplinary Anthology*, New Haven: Yale University Press.

Admassie, Yeraswork. 2000. *Twenty Years to Nowhere – Property Rights, Land Management and Conservation in Ethiopia*. Red Sea Press, Lawrenceville, NJ.

Africa Report, Ethiopia raises US$350 million for mega dam, Africa Report, 13 September 2011. Available online at http://nazret.com/blog/index.php/2011/09/13/ethiopia-raises-us-350-million-for-mega-dam (accessed 10 July 2016).

African Agriculture. Available online at http://africanagriculture.blogspot.com/2007/12/ethiopian-floriculture-continues-to.html (accessed 16 December 2007).

African Resource Working Group. 2008. Environmental and social impacts of the proposed Gibe III hydroelectric project in Ethiopia's lower Omo River Basin. Available online at www.forestpeoples.org/sites/fpp/files/publication/2010/08/ethiopiahydroelecimpactsmay08eng.pdf (accessed 20 August 2012).

Agarwal, Anil, and Sunita Narain. 1989. Towards green villages: a strategy for environmentally sound and participatory development. Quoted in Richard Douthwaite. 1999. *The Growth Illusion: How Economic Growth Has Enriched the Few, Impoverished the Many and Endangered the Planet*, Gabriola Island, Canada, p. 259.

Ahmed, Abdalla A., and Hamid A. E. Ismael. 2008. Sediment in the Nile River system, Khartoum: UNESCO International Sediment Initiative. Available online at www.irtces.org/isi/isi_document/Sediment%20in%20the%20Nile%20River%20System.pdf (accessed 17 January 2012).

Alemayehu, Tamiru. 1993. Preliminary analysis of the availability of groundwater in Ethiopia. *Ethiopian Journal of Science* 16(2): 43–59.

Alemayehu, Tamiru. 2000. Water pollution by natural inorganic chemicals in the central part of the main Ethiopian rift. *Ethiopian Journal of Science* 23, 2: 197–214.

Alemayehu, Tamiru. 2001. The impact of uncontrolled waste disposal on surface water quality in Addis Ababa. *Ethiopian Journal of Science* 24(1): 93–104.

Alemayehu, Tamiru. 2006. *Ground Water Occurrence in Ethiopia*, Addis Ababa: UNESCO, p. 75. Availabe online at www.eah.org.et/docs/Ethiopian%20groundwater-Tamiru.pdf (accessed 29 November 2010).

Amede, Tilahun, Takele Belachew, and Endrias Geta. 2001. *Reversing the Degradation of Arable Land in the Ethiopian Highlands*. Managing Africa's Soils No. 23, IIED, p. 4.

Amsalu, Aklilu, and Alebachew Adem. 2009a. *Assessment of Climate Change-induced Hazards, Impacts and Responses in Southern Lowlands of Ethiopia, Forum for Social Studies Research*. Report No. 4, Addis Ababa: Forum for Social Studies, p. 23.

Amsalu, Aklilu, and Alebachew Adem. 2009b. *Assessment of Climate Change-Induced Hazards, Impacts and Responses in Southern Lowlands of Ethiopia*. Research Report No. 4, Addis Ababa: Forum for Social Studies, p. 55.

Angassa, A. and Gufu Oba. 2009. Bush encroachment control demonstrations in southern Ethiopia: 1. Woody species survival strategies with implications for herder land management. *African Journal of Ecology*, 47: 63–76.

Appelgren, Bo, Wulf Klohn, and Undala Alam. 2000. *Water and Agriculture in the Nile Basin, Nile Basin*. Initiative Report to ICCCON, Food and Agriculture Organization, Rome, p. 3.

Arsano, Y., and I. Tamrat. 2005. Ethiopia and the Eastern Nile Basin. *Aquatic Sciences* 67: 15–27.

Asfaw, Desta. 1993. *Large-Scale Agricultural Development and Survival Issues among Pastoralists in the Awash Valley*. Conference on Pastoralism in Ethiopia, 4–6 February, Addis Ababa.

Asfaw, Gedion (ed.). 2003. *Environment and Environmental Change in Ethiopia*, Addis Ababa: Forum for Social Studies.

Ashenafi, Zelalem Tefra. 1994. *Management Plan of Omo National Park*, Addis Ababa: Ethiopian Wildlife Conservation Organization, pp. 36–7.

Awas, Tesfaye, Tamrat Bekele, and Sebsebe Demissew. 2001. An ecological study of the vegetation of Gambela Region, southwestern Ethiopia. *SINET: Ethiopian Journal of Science* 24, 2: 213–28.

Awulachew, Seleshi Bekele, and Mekonnen Ayana. 2010. Performance of irrigation: an assessment at different scales in Ethiopia, *Experimental Agriculture* 47, 1: 57–69.

Ayenew, Tenalem. 2001. Recent changes in the level of Lake Abiyata, Central Main Ethiopian Rift. *Journal of Hydrological Sciences* 47, 3: 493–503.

Ayenew, Tenalem. 2003. Evapotranspiration estimation using thematic mapper spectral satellite data in Ethiopian rift and adjacent highlands. *Journal of Hydrology* 279: 83–93.

Ayenew, Tenalem. 2004. Environmental implications of changes in the levels of lakes in the Ethiopian Rift since 1970. *Regional Environmental Change* 4: 12–204.

Ayenew, Tenalem. 2007. Some improper water resources utilization practices and environmental problems in the Ethiopian Rift. *African Water Journal* 1, 1: 81–100.

Bai, Z. G., D. L. Dent, L. Olsson, and M. E. Schaepman. 2008. *Global Assessment of Land Degradation and Improvement: 1. Identification by Remote Sensing*, Wageningen, The Netherlands, World Soil Information, Report 2008/01ISRIC, p. 25.

Baker, Richard St. Barbe. 1964. Some reflections on trees and forests for Ethiopia, *Ethiopia Observer* 8, 2: 189–92.

Barber, R. 1984. *An Assessment of the Dominant Soil Degradation Process in the Ethiopian Highlands: Their Impacts and Hazards*, Addis Ababa: Ministry of Agriculture, Land Use Planning and Regulatory Department.

Bassett, Thomas J. and Donald Crummey (eds.). 2003. *African Savannas: Global Narratives and Local Knowledge of Environmental Change*, Oxford: James Currey.

Bassett, Thomas J., and Donald. E. Crummey (eds.). 1993. *Land in African Agrarian Systems*, Madison: The University of Wisconsin Press, p. 256; pp. 247–73.

BBC Video, The Destruction of Ethiopia's Once Beautiful Lake Koka, 21 February 2009. Available online at http://nazret.com/blog/index.php?title=the_destruction_of_ethiopia_s_once_beaut_1&more=1&c=1&tb=1&pb=1 (accessed 21 December 2009).

Beadle, L. C. 1981. *The Inland Waters of Tropical Africa*, London: Longman Group Limited. Cited in Emily Peck, *Lake Turkana Conservation Science Program-WWF-US*. Available online at www.feow.org/ecoregion_details.php?eco=530 (accessed 7 December 2010).

Bekele, Melaku. 1998. The Ethiopian forest from ancient time to 1900: a brief account. *Walia* 19: 3–9.

Bekerie, Ayele. 1997. *Ethiopic: An African Writing System*, Lawrenceville, NJ: Red Sea Press.

Beltran, Javiar. 2000. Case study 6: Semien Mountain National Park, Ethiopia. In: *Indigenous and Traditional Peoples and Protected Areas: Principles, Guidelines, and Case Studies*, World Commission on Protected Areas (WCPA), p. 79.

Berhanu, Girma, and Taddese Yiberta. 1984. *The Nomadic Areas of Ethiopia Study Report: Part 3 – The Socio-economic Aspects*, United Nations Development Program, Addis Ababa, pp. 11–2.

Bewket, Woldeamlak. 2005, Biofuel consumption, household level tree planting and its implications for environmental management in the northwestern highlands of Ethiopia, *EASSRR* 21, 1: 19–38.

Beyene, Desta. 1988. Soil fertility research on some Ethiopian vertisols. In: *Proceedings of a Conference on Management of Vertisols in Sub-Saharan Africa held at ILCA*, Addis Ababa, 31 August–4 September 1987, pp. 223–31; Jutzi, p. 42.

Beyene, Fekadu. 2009. Exploring incentives for rangeland enclosures among pastoral and agro-pastoral households in eastern Ethiopia. *Global Environmental Change* 19, 4: 494–502.

Binns, Toney, Alan Dixon, and Etienne Nel. 2012. *Africa: Diversity and Development*, New York: Routledge.

BirdLife International. 2008. *BirdLife's Online World Bird Database: The Site for Bird Conservation*. Version 2.1. Cambridge, UK: BirdLife International. Available online at www.birdlife.org (accessed 16 August 2010).

BirdLife International. 2011. *Important Bird Areas factsheet: Babille Elephant Sanctuary*. Available online at www.birdlife.org (accessed 18 March 2011).

Blonk, Matthijs. 2011. Indigenous peoples versus the 'business model': why African Parks Networks is pulling out of Ethiopia, *Ecologie and Ontwikking*, a magazine of IUCN-NL, September 2008. Available online at www.matthijsblonk.nl/paginas/AfricanParksEthiopieEng.htm (accessed 19 March 2011).

Bolton, M. 1969. Rift Valley Ecological Survey, UNESCO, p. 14.

Bondestam, Lars. 1974. People and Capitalism in Northeastern Lowlands of Ethiopia, *Journal of Modern African Studies* 12, 3: 423–39.

Booming floriculture industry fuels Ethiopian economic growth. *The Guardian*, 14 February 2008. http://africanagriculture.blogspot.com/2008/02/booming-floriculture-industry-fuels.html.

Boserup, E. 1981. *Population and Technology*. Oxford: Basil Blackwell.

Bourn, David. 2002. *Farming in Tsetse Controlled Areas of Eastern Africa Ethiopia National Component: Farming Systems and Natural Resource Management*. Short Term Technical Assistance Consultancy Report, Project 7 ACP ET086, Addis Ababa, Ministry of Agriculture, p. 7.

Braimola, Ademola K., and Paul L. G. Vlek (eds.). *Land Use and Soil Resources*, Stockholm: Swedish Academy of Sciences, p. 73; pp. 73–100.

Briggs, Philip. 2009. *Ethiopia: The Bradt Travel Guide*, Guilford, CT: The Globe Pequot Press, pp. 426–7.

Brink, M. and G. Belay. 2006. *Plant Resources of Tropical Africa 1: Cereals and Pulses*, Wageningen: PROTA Foundations, The Netherlands.

British Geological Survey (BGS). 2001. *Groundwater Quality: Ethiopia, British Geological Survey*. Available online at www.wateraid.org/documents/plugin_documents/ethiopiagw.pdf (accessed 29 November 2010).

Bruce, John W., Allan Hoben, and Dessalegn Rahmato. 1994. *After the Derg: An Assessment of Rural Land Tenure Issues in Ethiopia*, Madison: Land Tenure Center, University of Wisconsin at Madison.

Butzer, Karl. 1981. Rise and fall of Axum, Ethiopia: a geo-archaeological interpretation. *American Antiquity* 46, 3: 471–495.

Camerapix Publishers International. 1995. *African Wildlife Safaris*, Nairobi, Kenya: Camerapix, p. 134.

Campbell, John. 1991. Land or peasants? The dilemma confronting Ethiopian resource conservation. *African Affairs*, 90: 5–21.

Canadian Hunger Foundation International (CHF). 2006. Grassroots conflict assessment of the Somali region, Ethiopia, August. Available online at www.chfhq.org/files/3707_file_Somali_Region_Assessment_8.4.06.pdf (accessed 17 November 2010).

Carr, C. J. 1998. Patterns of vegetation along the Omo River in Southwest Ethiopia. *Plant Ecology* 135:135–163.

Cascao, A. E. 2008. Ethiopia – Challenges to Egyptian hegemony in the Nile Basin. *Water Policy* 10: 13–28.

Central Statistical Agency (CSA). 1984. Ethiopia 1984: Population and Housing Preliminary Report, 1, 1, Addis Ababa: CSA.

Central Statistical Agency (CSA). 2011. *Statistical Abstract 2011/2012*, Addis Ababa: CSA, p. 7. Available online at http://ewnhs.org.et/wp-content/uploads/downloads/2011/03/Biodiversity-Hotspots-of-Ethiopia.pdf (accessed 4 November 2014).

Centre for Environmental Economics and Policy in Africa (CEEPA). 2006. *Climate Change and African Agriculture: Assessing the Impact of Climate Change on the Water Resources of the Lake Tana Sub-basin Using the Watbal Model*, Policy Note No.30, CEEPA: University of Pretoria, South Africa, p. 2. Available at www.ceepa.co.za/docs/POLICY%20NOTE%2030.pdf (accessed 8 December 2010).

Chala, Zelalem T. 2010. *Economic Significance of Selective Export Promotion on Poverty Reduction and Inter-Industry Growth of Ethiopia*. PhD dissertation, Virginia Polytechnic Institute and State University, p. 13.

Chernet, Tesfaye, Y. Travi, and V. Valles. 2001. Mechanism of degradation of the quality of natural waters in the lakes region of the Ethiopian rift valley. *Water Resources* 35: 2819–32.

Clay, Jason W. and Bonnie K. Holcomb. 1986. *Politics and the Ethiopian Famine 1984–1985*, New Brunswick, NJ: Transaction Books.

Clay, Jason W., Sandra Steingraber, and Peter Nigel. 1988. *The Spoils of Famine*, Cambridge, MA: Cultural Survival, Inc.

Clay, Jason. 2004. *World Agriculture and the Environment*, London: Island Press.

Conway, D. 1997. A water balance model for the upper Blue Nile basin in Ethiopia. *Hydrological Sciences Journal* 42: 256–86.

Conway, D. 2000. The climate and hydrology of the upper Blue Nile River. *The Geographical Journal* 166: 49–62.

Coppock, D. L. 1994. The Borana Plateau of Southern Ethiopia: Synthesis of Pastoral Research, Development and Change, 1980–91, Addis Ababa: International Livestock Research Institute (ILRI).

Cossins, N. J. and M. Upton. 1987. The Borana pastoral system of southern Ethiopia. *Agricultural System* 25: 199–218.

Coulter, G. W., B. R. Allanson, et al. 1986. Unique qualities and special problems of the African Great Lakes. *Environmental Biology of Fishes* 17, 3: 161–83.

CSA, 2010. Livestock and Livestock Characteristics, Agricultural Sample Survey 2009, Statistical Bulletin 468, Addis Ababa: CSA, p. 39.

CSA. 1995. The 1994 Population and Housing Census, Addis Ababa: CSA.

CSA. 2009. Report on Land Utilization, Volume IV, Statistical Bulletin No. 446, Addis Ababa: CSA, Table 1, p. 13.

CSA. 2010. Statistical Abstract 2009, Addis Ababa: CSA, p. 44.

Cushion, Elizabeth, Adrian Whiteman, and Gerhard Dieterle. 2010. *Bio-energy Development: Issues and Impacts for Poverty and Natural Resources Management*, Washington, DC: The World Bank.

Daly, Herman E. 1996. Sustainable growth? No thank you. In: *The Case against the Global Economy and For a Turn Toward the Local*, Mander, Jerry, and Edward Goldsmith (eds.), San Francisco: Sierra Club Book, p. 196.

Darbyshire, Iain, Henry Lamb, and Mohammed Umer. 2003. Forest clearance and regrowth in northern Ethiopia during the last 3000 years. *The Holocene* 13(4): 537–46.

Dasmann, Raymond F., John P. Milton, and Peter H. Freeman. 1973. *Ecological Principles for Economic Development*, New York: John Wiley, p. 204.

Davison, William. 2010. Ethiopia relocates 150,000 people in eastern Somali region in five months. *Bloomberg News*, November 29, 2010. Available online at www.bloomberg.com/news/2010-11-29/ethiopia-relocates-150-000-people-in-eastern-somali-regionin-five-months.html (accessed 4 December 2010).

De Blij, Harm J. and Peter O. Muller. 2008. *Geography: Realms, Regions, and Concepts*, NY: Wiley.

Debele, Berhanu. 1989. The role of land use planning in Ethiopia's national food strategy, In: Towards a Food and Nutrition Strategy in Ethiopia. *The Proceedings of the National Workshop on Food Strategies for Ethiopia*, Haromaya University, 8–12 December, 1986, pp. 180–205.

Demeke, Mulat, Ali Said, and T. S. Jayne. 1997. Promoting fertilizer use in ethiopia: the implication of improving grain market performance, input market efficiency, and farm management, Working Paper 5, Grain Market Research Project, Ministry of Economic Development and Cooperation, Addis Ababa.

Demeke, Yirmed, and Negusu Aklilu. 2008. Alarm bell for biofuel development in Ethiopia: the case of Babille Elephant Sanctuary. In: *Agrofuel Development in Ethiopia: Rhetoric, Reality and Recommendations*, Heckett, T., and Negusu Aklilu (eds.), Addis Ababa: Forum for Environment in partnership with Horn of Africa Regional Environmental Center/Network, pp. 83–113.

Demissew, Sebsebe. 1988. The floristic composition of Menagesha state Forest and the need to conserve such forests in Ethiopia. *Mountain Research and Development* 8: 243–7.

Deribe, Shewaye. 2008. Wetland management aspects in Ethiopia: Situation analysis. In: *The Proceedings of the National Stakeholders' Workshop on Creating National Commitment for Wetland Policy and Strategy Development in Ethiopia*, 7–8 August, Addis Ababa, Ethiopia, pp. 14–27. Available online at www.ewnra.org.et/files/Proceedings%20of%20 the%20national%20stakeholders%27%20workshop.pdf (accessed 1 December 2010).

Desta, Solomon and Layne D. Coppock. 2004. Pastoralism under pressure: tracking system change in southern Ethiopia. Human Ecology 32(4): 465–86.

Desta, Zerihun. 2000. Challenges and opportunities of Ethiopian wetlands: the case of Lake Awassa and its feeders. In: *Proceedings of a Seminar on the Resources and Status of Ethiopia's Wetlands*, Yilma D. Abebe and Kim Geheb (eds.), The World Conservation Union (IUCN), pp. 67–74.

Di Paola, G. M. and Getahun Demissie. 1979. Geothermal energy: an inexhaustible resource of great economic importance for Ethiopia. *SINET: Ethiopian Journal of Science* 2(2): 87–110.

Diamond, Jared. 2011. *Collapse: How Societies Choose to Fail or Succeed*, New York: Penguin Books.

Dixon, Alan B. 2003. *Indigenous Management of Wetlands: Experiences in Ethiopia*. London: Ashgate Publishing.

Dixon, Alan B. 2005. Wetland sustainability and the evolution of indigenous knowledge in Ethiopia. *The Geographical Journal* 171: 306–23.

Dixon, Alan, and Adrian Wood. 2001. *Sustainable Wetland Management for Food Security and Rural Livelihoods in South-west Ethiopia: The Interaction of Local Knowledge and Institutions, Government Policies and Globalization*. Paper Prepared for Presentation at the Seminaire sur L'amenagement des zones marecageuses du Rwanda, 5–8 June, National University of Rwanda and the University of Huddersfield and Wetland Action.

Douthwaite, Richard. 1999. *The Growth Illusion: How Economic Growth Has Enriched the Few, Impoverished the Many and Endangered the Planet*, Gabriola Island, Canada.

Dubale, Paulos. 2001. Soil and water resources and degradation factors affecting productivity in Ethiopian highland agro-ecosystems. *Northeast African Studies* 8, 1: 27–52.

Dufey, Vermeulen. 2007. *Biofuels Strategic Choices for Commodity Dependent Developing Countries*. Common Fund for Commodities (CFC), Amsterdam, The Netherlands.

Eguavoen, Irit. 2009. *The Acquisition of Water Storage Facilities in the Abbay River Basin, Ethiopia*, Working Paper Series 38, Center for Development Research (ZEF), University of Bonn.

Egziabher, Tewolde Berhan Gebre. 1986. Ethiopian vegetation – past, present and future. *SINET: Ethiopian Journal of Science* 9(suppl.): 1–13.

Egziabher, Tewolde Berhan Gebre. 1988. Vegetation and environment of the mountains of Ethiopia: implications for utilization and conservation. *Mountain Research and Development* 8: 211–6.

Egziabher, Tewolde Berhan Gebre. 1991. Management of mountain environments and genetic erosion in tropical mountain systems: The Ethiopian example. *Mountain Research and Development* 11, 3: 225–30.

Eide, A. 2008. *The Right to Food and the Impact of Liquid Biofuels (Agrofuels)*. Right to Food Studies. Rome: Food and Agriculture Organization.

Elias, Eyasu, and Feyera Abdi. 2010. *Putting Pastoralists on the Policy Agency: Land Alienation in Southern Ethiopia, International Institute for Environment and Development (IIED)*. Gate Keeper Series # 145, July, p. 10. Available online at www .iied.org/pubs/pdfs/14599IIED.pdf (accessed 16 November 2010).

Ellis, S. and A. Miller. 1995. *Soils and Environment*. New York: Routledge.

Engels, J. M., J. G. Hawkes, and M. Worede. 1991. *Plant Genetic Resources of Ethiopia*, Cambridge: Cambridge University Press.

Engstrom, L. 2009. *Liquid Biofuels – Opportunities and Challenges in Developing Countries*: A Summary Report from SIDA's Helpdesk for Environmental Assessment, May.

Environmental Protection Agency (Ethiopia). 1997. *A Draft Proposal for the Establishment of National Desertification Fund of Ethiopia*, Addis Ababa: EPA, p. 3.

Environmental Protection Authority. 1998. *National Action Program to Combat Desertification*, Addis Ababa: EPA, p. 14.

Environmental, labor concerns grow over Ethiopian floriculture industry, *The East African*, 19 February 2000. Available online at http://africanagriculture.blogspot.com/2008/02/environmental-labour-concerns-grow-over.html.

ERATA News, 2 April 2011. Ethiopia lays foundation for Africa's largest dam. Available online at www.ertagov.com/erta/erta-news-archive/38-erta-tv-hot-newsaddis-ababa-ethiopia/574-ethiopia-lays-foundation-for-africas-biggest-dam.html (accessed 1 September 2012).

Ethiopia Electric Power Corporation. 2009. Available online at www.eepco.gov.et (accessed 20 December 2009); UNESCO, p. 59.

Ethiopia's Rush to Build Mega Dams Sparks Protests. *The Guardian* (UK), 25 March 2010.

Ethiopian Mapping Authority. 1988. *National Atlas of Ethiopia*, Addis Ababa: Ethiopian Mapping Authority, p. 8.

Ethiopian Mapping Authority. 1988. *National Atlas of Ethiopia*, Addis Ababa, Ethiopian Mapping Authority, p. 60). Fourteen of the birds and eleven of the mammals endemic to the country are found in this park.

Ethiopian NAPA. 2007. *Climate Change National Adaptation Program of Action*, Addis Ababa, Ethiopia

Ethiopian Science and Technology Commission. 2004. *Proposal for Organizing a National Roundtable on Sustainable Consumption and Production in Akaki River Basin*, Addis Ababa. Available online at www.gadaa.com/AkakiRiverRoundtable.pdf (accessed 22 December 2010).

Ethiopian Valleys Development Studies Authority (EVDSA). 1989. *Master Plan for the Development of Surface Water Resources in the Awash Basin*. Final Report, volume 9, Addis Ababa, p. 9.

Ethiopian Valleys Development Studies Authority (EVDSA). 1992. *A Review of Water Resources of Ethiopia*, Addis Ababa: EVDSA, p. 38.

Ethiopian Wildlife and Natural History Society (EWNHS). 1996. *Important Bird Areas of Ethiopia*, Addis Ababa: EWNHS and Birdlife International, p. 111.

Ethiopian Wildlife and Natural History Society (EWNHS). 1996. *Important Bird Areas of Ethiopia: A First Inventory*, Addis Ababa: EWNHS, p. 35.

Ethiopian Wildlife Conservation Authority (EWCA). 2007. *Bale Mountains National Park: General Management Plan 2007, 2017*, Addis Ababa: EWCA, p. 61.

EVDSA. 1989b. *Master Plan for the Development of Surface Water Resources in the Awash Basin*, Volume 4, Annex A: Climate and Hydrology, Addis Ababa: EVDSA, pp. 1–3.

EWCO. 1991. *Semien Mountains National Park*, Situation Report to UNESCO, Ethiopia.

Eyasu Elias. 2008. *Pastoralists in Southern Ethiopia: Dispossession, Access to Resources and Dialogue with Policy Makers*. Drylands Coordinating Group Report No. 53, p. 1. Available online at www.drylands-group.org/Articles/1451.html (accessed 9 November 2010).

Ezra, Markos. 1997. *Demographic Responses to Ecological Degradation and Food Insecurity: Drought Prone Areas in Northern Ethiopia.* PhD dissertation, The Netherlands Graduate School of Research and Demography, p. 63; WBISPP, p. 53.

FAO. 1997. *Irrigation Potential in Africa: A Basin Approach*, Rome: FAO, p. 67.

Federal Democratic Republic of Ethiopia. 2010. *Growth and Transformation Plan, 2010/11–2014/15*, Addis Ababa, p. 9.

Fenta, Yichilal. 1995. *Wildlife Resources in Ethiopia with Particular Reference to Mammals.* Geography senior thesis, Addis Ababa University, p. 17. 36 Fanuel Kebede. 1999b, p. 7.

Fisseha, Messele. 2003. Water resources policy and river basin development as related to wetlands, In: *Wetlands of Ethiopia. Proceedings of a Seminar on the Resources and Status of Ethiopia's Wetlands*, Abebe, Yilma D., and Kim Geheb (eds.), Nairobi, IUCN Eastern Africa Regional Office, p. 78.

Fitzgerald, M. 1981. Sibiloi: The remotest park in Kenya. *Africana* 8, 4: 22.

Food and Agriculture Organization of the United Nations (FAO). 2005. *Ethiopia: Irrigation in Africa in Figures – Aquastat Survey 2005*, Rome: FAO.

Friis, Ib, Sebsebe Demissew, and Paulo van Breugel. 2011. *Atlas of the Potential Vegetation of Ethiopia*, Addis Ababa: Addis Ababa University Press.

Galperin, Georgi. 1978. *Ethiopia: Population, Resources, Economy*, Moscow: Progress Publishers.

Ganesamurthy, V. S. 2011. *Environmental Status and Policy in India*, New Delhi: New Century Publications, p. xxv.

Gashaw, Menassie and Masresha Fetene. 1996. Plant communities of the Afroalpine vegetation of Sanetti Plateau, Bale Mountains, Ethiopia. *SINET, Ethiopian Journal of Science* 19, 1: 65–86.

Gebre-Egziabher, Tewolde Berhan. 1991. Diversity of Ethiopian flora. In: *Plant Genetic Resources of Ethiopia*, J. M. M. Englles, J. G. Hawkes, and Melaku Worede (eds.), Cambridge: Cambridge University Press, pp. 75–81.

Gebreselassie, Samuel. 2006. *Land, Land Policy and Smallholder Agriculture in Ethiopia: Options and Scenarios.* Paper prepared for the Future Agricultures Consortium meeting at the Institute of Development Studies, 20–22 March 2006, Addis Ababa: Institute of Development Research, p. 13.

Gemessa, Kejela, Bezabih Emana, and Waktole Tiki. 2006. *Livelihood Diversification in Borana Pastoral Communities of Ethiopia – Prospects and Challenges*, Addis Ababa. Available online at www.ilri.org/Link/Publications/Publications/Theme%201/ Pastoral%20conference/Papers/Gemtessa%20Livelihood%20Diversification%20 of%20the%20Pastoral%20Communities%20of%20Borena.pdf (accessed 16 November 2010).

Geological Survey of Ethiopia, Current Exploration and Mining. Available online at www .telecom.net.et/~walta/profile/html/current_status.html (accessed 13 December 2010).

Getachew, Kassa Negussie. 2001. *Among the Pastoral Afar in Ethiopia: Tradition, Continuity and Socio-economic Change.* Utricht, The Netherlands.

Gete, W. No date. *Natural Resources, Poverty, and Conflict in Sub-Saharan Africa: Evidence from Ethiopia*, pp. 18–25.

Geza, Mengistu. 1999. Harnessing techniques and work performance of draught horses in Ethiopia. In: *Meeting the Challenges of Animal Traction.* A resource book of the Animal Traction Network for Eastern and Southern Africa (ATNESA), Starkey, P., and P. Kaumbutho (eds.), Harare, Zimbabwe: Intermediate Technology Publications, pp. 144–7; p. 144.

Ghimire, Krishna B., and Michael P. Pimbert. 1997. Social change and conservation: an overview of issues and concepts. In: *Social Change and Conservation: Environmental Politics and Impacts of National Parks and Protected Areas*, Ghimire, Krishna B., and Michael P. Pimbert (eds.), London: Earthscan, pp. 1–45; p. 21.

Gibbons, Ann. 2009. *Ancient Skeleton May Rewrite Earliest Chapter of Human Evolution, Science Now*, 1 October.

Gibrnachen, Zemenawi, vol. 3, No. 1, Yekatit 2006 E.C. (The Ministry of Agriculture Bulletin), p. 14.

Gizaw, Berhanu. 1996. The origin of high bicarbonate and fluoride in waters of the main Ethiopian Rift Valley, East African Rift system, *Journal of African Earth Sciences* 22(4): 391–402.

Glantz, Michael H. (ed.). 1987. *Drought and Hunger in Africa: Denying Famine a Future*, Cambridge: Cambridge University Press.

Gordon, Francis L. 2000. *Ethiopia, Eritrea and Djibouti*, London: Lonely Planet Publications.

Goudie, Andrew S. 1975. Former lake levels and climatic change in the Rift Valley of southern Ethiopia. *Geographical Journal* 141: 177–94.

Goudie, Andrew. 2006. *The Human Impact on the Natural Environment*. Cambridge, MA: Cambridge University Press.

Graham, Jane. n.d. *National Parks in Ethiopia's Southern Region*, Awassa, Agricultural Bureau, Southern, Nations, Nationalities and Peoples Region.

Green, Adam W. M. 1990. *Development: Environment and Sustainability in the Third World*, New York: Routledge, p. 186.

Greste, Peter. 2009. The dam that divides Ethiopians. *BBC News*, 26 March 2009. Available online at http://news.bbc.co.uk/2/hi/africa/7959444.stm (accessed 2 December 2010).

Grove, A. T., F. A. Street, and A. S. Goudie. 1975. Former lake levels and climatic change in the Rift Valley of southern Ethiopia. *Geographical Journal* 141: 177–94.

Guariso, G. and D. Whittington. 1987. Implication of Ethiopian water development for Egypt and Sudan. *Water Resources Development* 3(2): 105–14.

Gudeta, Zerihun. 2009. How successful the Agricultural Development-Led Industrialization Strategy (ADLI) will be leaving the existing landholding system intact? A major constraint for the realization of ADLI's target. *Ethiopian e-Journal for Research and Innovation Foresight* 1, 1 (December), p. 3

Haberson, John W. 1978. Territorial and development politics in the Horn of Africa: the Afar of the Awash Valley. *African Affairs* 77, 309: 479–98.

Hagos, Fitsum, G. Makombe, R. E. Namara, and Seleshi Bekele Awulachew. 2009. *Importance of Irrigated Agriculture to the Ethiopian Economy: Capturing the Direct Net Benefits of Irrigation*. Colombo, Sri Lanka: International Water Management Institute, IWMI Research Report 128, p. 7.

Haileselassie, Ayenew. 2004. Ethiopia's struggle over land reform, *World Press Review*, Vol. 51, No. 4. Available online at www.worldpress.org/Africa/1839.cfm (accessed 17 August 2011).

Haileslassie, Amare, Fitsum Hagos, Everisto Mapedza, Claudia Sadoff, Seleshi Bekele Awulachew, Solomon Gebreselassie, and Don Peden. 2008. *Institutional Settings and Livelihood Strategies in the Blue Nile Basin: Implications for Upstream/ Downstream Linkages*, Working Paper 132, International Livestock Research Institute (ILRI) and International Water Management Institute, p. 22. Available online at www .indiaenvironmentportal.org.in/files/WOR132.pdf (accessed 5 October 2010).

Halcrow, Sir William, et al. 1989. *Master Plan for the Development of Surface water Resources in the Awash Basin, Ethiopian Valleys Development Authority*, Vol. 8, Annex J: Irrigation and Drainage.

Halfman, J. D., and T. C. Johnson. 1988. High resolution record of cyclic climate change during the past 4 ka from Lake Turkana, Kenya, *Geology*, 16, 496–500; Cited in Velpuri, N. M., and G. B. Senay. 2012. Assessing the potential hydrological impact of the Gibe III Dam on Lake Turkana water level using multi-source satellite data. *Hydrology and Earth System Sciences* 9: 2987–3027.

Hancock, Graham, Richard Pankhurst, and Duncan Willetts. 1993. *Under Ethiopian Sky*, London: H & L Communication.

Handleman, Howard. 2011. *The Challenge of Third World Development*. London: Longman.

Harrison, Paul. 1987. *The Greening of Africa: Breaking through in the Battle for Land and Food*. London: Penguin.

Harvey, Celia A. and David Pimentel. 1996. Effects of soil and wood depletion on biodiversity. *Biodiversity and Conservation* 5: 1121–30.

Hathaway, Terri. 2008. What cost Ethiopia's dam boom? *International Rivers*, February: 4–26.

Hedberg, O. 1986. The Afroalpine flora of Ethiopia. *SINET: Ethiopian Journal of Science*. 9: 105–10.

Helland, J. 1980. *An Analysis of Afar Pastoralism in the Northeastern Rangeland of Ethiopia*, African Savannah Studies, Occasional paper No. 20.

Helland, Johan. 1997. Development interventions and pastoral dynamics in southern Ethiopia. In: *Pastoralists, Ethnicity and the State in Ethiopia*, Hogg, Richard, (ed.), London: Haan Publishing, pp. 62–4.

Helland, Johan. 1999. Land Alienation in Borana: some land tenure issues in a pastoral context in Ethiopia. *Eastern Africa Social Science Research Review* 14(2): 47–65.

Hengsdijk, Huib, and Herco Jansen. 2006. *Agricultural Development in the Central Ethiopian Rift Valley: A Desk-Study on Water-Related Issues and Knowledge to Support a Policy Dialogue*, Plant Research International B.V., Wageningen.

Higgins-Zogib, Liza, Nigel Dudley, Stephanie Mansourian, and Surin Suksuwan. 2010. Safety net: protected areas contributing to human well-being. In: *Arguments for Protected Areas: Multiple Benefits for Conservation and Use*, Stolton, Sue, and Nigel Dudley (eds.), London: Earthscan, p. 128.

Hillel, David. 2008. *Soil in the Environment: Crucible of Terrestrial Life*, Boston: Academic Press.

Hillman, J. C. 1983. *Ethiopia Compendium of Wildlife Conservation Information*, Vol. 1, Addis Ababa: Ethiopian Wildlife Conservation Organization, p. 9.

Hillman, J. C. 1986. Conservation in Bale Mountains National Park, Ethiopia. *Oryx* 20: 89.

Hillman, J. C. 1993. *Ethiopia: Compendium of Wildlife Conservation Information*, The Wildlife Conservation Society International, New York Zoological Society, New York, Ethiopian Wildlife Conservation Organization.

Hoben, Allan. 1995. Paradigms and politics: the cultural construction of environmental policy in Ethiopia. *World Development* 23(6): 1007–21.

Hogg, Richard (ed.). 1987. *Pastoralists, Ethnicity and the State in Ethiopia*, London: Haan Publishing.

Homann, Sabine, B. Rischkowsky, and J. Steinbach. 2007. The effect of development interventions on the use of indigenous range management strategies in the Borana lowlands in Ethiopia. *Land Degradation and Development* 19: 368–87.

Homann, Sabine. 2005. *Indigenous Knowledge of Borana Pastoralists in Batural Resource Management: A Case Study from Southern Ethiopia*, Cuvillier Verlag, Gottingen, p. 54.

Honey, Martha. 2008. *Ecotourism and Sustainable Development*, London: Island Press, p. 13.

Hopson, A. J., 1982. *Lake Turkana. A Report on the Findings of the Lake Turkana Project 1972–1975*, Volumes 1–6, London, UK: Overseas Development Administration.

Horvath, Ronald J. 1968. Addis Ababa's eucalyptus forest. *Journal of Ethiopian Studies*, 6, 1: 13–9.

Horvath, Ronald J. 1968. Towns in Ethiopia. *Erkunde* 22: 42–8.

Howell, P. P. and J. A. Allan. 1994. *The Nile: Managing a Scarce Resource*. Cambridge: Cambridge University Press.

Huffnagel, H. P. 1961. *Agriculture in Ethiopia*, Rome: Food and Agriculture Organization.

Hughes F. M. 1988. The ecology of Africa flood plain forests in semi-arid and arid-zones review. *Journal of Biogeography* 15:127–40.

Hughes, R. H., and J. S. Hughes. 1992. *A Directory of African Wetlands*, Gland, Switzerland, Nairobi, Kenya, and Cambridge, UK: IUCN and UNEP.

Hundessa, Tesfaye. 1996. Utilization of wildlife in Ethiopia. *Walia: Journal of the Ethiopian Wildlife and Natural History Society* 17: 3–9.

Hurni, H. 1986. *Management plan Semien Mountains National Park and Surrounding Rural Area*. UNESCO, World Heritage Committee and Wildlife Conservation Organization.

Hurni, H. and B. Messerli. 1981. Mountain research for conservation and development in Semien, Ethiopia. *Mountain Research and Development* 1(1): 49–54.

Hurni, Hans and Eva Ludi. 1994. *Reconciling Conservation with Sustainable Development*, Center for Development and Environment, University of Berne, Switzerland. Available online at www.cde.unibe.ch/Tools/pdf/afr22_part1.pdf.

Huxley, Gille A., L. Swift Monod, and E. B. Worthington. 1963. *The Conservation of Nature Natural Resources in Ethiopia*, Paris: UNESCO/NS/NR/47, p. 13.

Institute of Biodiversity. 2008. *Conservation, Ecosystems of Ethiopia*, Addis Ababa. Available online at www.ibc-et.org/ (accessed 6 January 2011).

International Labor organization (ILO). 2006. *Convention Concerning Indigenous and Tribal Peoples in Independent Countries*, Geneva: ILO. Available online at www.ilo. org/ilolex/cgi-lex/convde.pl?C169/ (accessed 19 March 2011).

International Rivers. 2009. Ethiopia's Gibe III Dam: Sowing Hunger and Conflict, Berkeley, CA.

IUCN. 1990. *Biodiversity in Sub-Saharan Africa and Its Islands: Conservation, Management, and Sustainable Use*, Occasional Papers of the IUCN Species Survival Commission, IUCN – The World Conservation Union, p. 32.

IUCN. 1997. *The Ethiopian Wolf: Status Survey and Conservation Action Plan*, IUCN/ SSC (Canid Special Group), Compiled and edited by Claudio Sillero-Zubiri and David Macdonald.

Jacobs, Michael J. and Catherine A. Schloeder. 2001. *Impact of Conflict on Biodiversity and Protected Areas in Ethiopia*, Washington, DC: World Wildlife Fund.

Jagger, P. and J. Pender. 2003. The role of trees in sustainable management of less favored lands: the case of eucalyptus in Ethiopia. *Forest Policy and Economics* 5: 83–95.

Jansen, Herco, Huib Hengsdijk, Dagnachew Legesse, Tenalem Ayenew, Petra Hellegers, and Petra Spliethoff. 2007. *Land and Water Resources Assessment in the Ethiopian Central Rift Valley: Project: Ecosystems for Water, Food and Economic Development in the Ethiopian Central Rift Valley*, Alterra-rapport. Wageningen, Alterra, p. 10.

Jansen, Herco, Huib Hensdijk, Dagnachew Legesse, Tenalem Ayenew, Petra Hellegers, and Peter Spliethoff. 2007. *Land and Water Resources Assessment in the Ethiopian Central Rift Valley*, Alterra-rapport 1587, Alterra, Wageningen, The Netherlands, p. 23.

Janssen, Kurt, Michael Harris, and Angela Penrose. 1987. *The Ethiopian Famine*, London: Zed Press.

Johnson, Thomas C., and John O. Malala. 2009. Lake Turkana and its link to the Nile. *Monographiae Biologicae* 89, 111: 287–304, p. 1.

Johnston, R., and M. McCartney. 2010. *Inventory of water storage types in the Blue Nile and Volta River Basins*. Colombo, Sri Lanka: International Water Management Institute, IWMI Working Paper 140, p. 7.

Joireman, S. 2000. *Property Rights and Political Development in Ethiopia and Eritrea*, East African Studies, Oxford: James Currey.

Jovanovic, D. 1985. Ethiopian interest in the division of the Nile River waters. *Water International* 10(2): 82–5.

Juma, Calestous. 2011. *The New Harvest: Agricultural Innovation in Africa*, Oxford: Oxford University Press.

Jutzi, Samuel C. 1989. The Ethiopian vertisols: A vast natural resource, but considerably underutilized. In: *First Natural Resources Conservation Conference, Natural Resources Degradation: A Challenge to Ethiopia*, February 1–8, 1989, Addis Ababa, pp. 41–45; p. 41.

Kagwanja, P. 2007. Calming the waters. The east African community and the conflict over the Nile resources. *Journal of Eastern African Studies* 1(3): 321–37.

Karyabwite, D. R. 2000. *Water Sharing in the Nile River Valley*, UNEP.

Kassa, Belay. 2003. Agricultural extension in Ethiopia: the case of participatory demonstration and training extension system. *Journal of Social Development in Africa* 18(1): 49–83.

Kebbede, Girma. 1992. *The State and Development in Ethiopia*, Atlantic Highlands, NJ: Humanities Press.

Kebbede, Girma. 2004. *Living with Urban Environmental Health Risks: The Case of Ethiopia*. London: Ashgate Publications.

Kebede, Fanuel, L. Berhanu, and P. D. Moehlman. 2007. *Distribution and Population Status of the African Wild Ass (Equus africanus) in Ethiopia*. Report to Saint Louis Zoo.

Kebede, Fanuel. 1999a. Equus africanus. In: *IUCN Red List of Threatened Species*. Available online at www.iucnredlist.org/apps/redlist/details/7949/0 (accessed 14 March 2010).

Kebede, Fanuel. 1999b. *Ecology and Conservation of the African Wild Ass (Equus africanus) in the Danakil, Ethiopia*. MSc thesis, Durrell Institute of Conservation and Biology, University of Kent at Canterbury, UK, p. 7.

Keiner, Marco. Towards a New Management Plan for the Semien Mountain National Park. Available online at http://e-collection.ethbib.ethz.ch/ecolpool/bericht/bericht_300.pdf (accessed 10 July 2016).

Keiner, Marco. Towards a New Management Plan for the Semien Mountain National Park. Available online at http://e-collection.ethbib.ethz.ch/ecol-pool/bericht/bericht_300.pdf (accessed 10 July 2016).

Keiner, Marco. Towards a New Management Plan for the Semien Mountain National Park. Available online at http://e-collection.library.ethz.ch/eserv/eth:26720/eth-26720-01.pdf (accessed 10 July 2016).

Kenya Wildlife Service. 2001. *Nomination Form for the Great Rift Valley Ecosystems Sites. Extension of Lake Turkana: South Island National Park*, Nairobi. Available online at www.unep-wcmc.org/sites/wh/pdf/Lake%20Turkana.pdf (accessed 7 December 2010).

Khalid, Kiran, Akaki River. Available online at www.gadaa.com/AkakiRiver.html (accessed 21 December 2010).

Kibret, Solomon, Matthew McCartney, Jonathan Lautze, and Gayathree Jayasinghe. 2009. *Malaria Transmission in the Vicinity of Impounded Water: Evidence from the Koka*

Reservoir, Ethiopia, IWMI Research Report 132, Colombo, Sri Lanka: International Water Management Institute, 2009. Available online at www.iwmi.cgiar.org/Publications/IWMI_Research_Reports/PDF/PUB132/RR132.pdf (accessed 22 December 2010).

Kidan, Woldegebriel Gebre. 1996. The Status of Mountain Nyala (*Trigelaphus buxtoni*) in the Bale Mountains National Park, 1986–1994. *Walia* 17, p. 27.

Kloos, H. and A. Lemma. 1977. Bilharziasis in the Awash Valley: epidemiological studies in the Nura Era, Abadir, Melka Sedi and Amibara. *Ethiopian Medical Journal* 15: 166–8.

Kloos, Helmut and Aynalem Adugna. 1989. The Ethiopian population: growth and distribution. *The Geographical Journal* 155(1): 33–51; p. 36.

Kloos, Helmut. 1982. Development, Drought, and Famine in the Awash Valley of Ethiopia. *African Studies Review* 25, 4: 21–48; pp. 5–7.

Koehn, Peter. 1979. Ethiopia: famine, food production, and changes in the legal order. *African Studies Review* 22(1): 51–71.

Kugelman, Michael, and Susan L. Levenstein. 2013. *The Global Farms Race: Land Grabs, Agricultural Investment, and the Scramble for Food Security*, London: Island Press, p. 71.

Laverenchenko, L. A., A. N. Milishnikov, V. M. Aniskin, A. A. Warshavsky, and Woldegabriel Gebrekidan. 1997. The genetic diversity of small mammals of the Bale Mountains, Ethiopia. *SINET: Ethiopian Journal of Science* 20(2): 213–33.

Lefort, Rene. 1983. *Ethiopia: An Heretical Revolution*. London: Zed Press.

Legesse, Dagnachew, and Tenalem Ayenew. 2006. Effect of improper water and land resource utilization on the central Main Ethiopian Rift lakes. *Quaternary International* 148, 1: 8–18; p. 14.

Leveque, C. 1997. *Biodiversity Dynamics and Conservation: The Freshwater Fish of Tropical Africa*, Cambridge, UK: Cambridge University Press.

Low, Pak Sum (ed.). 2006. *Climate Change and Africa*. Cambridge: Cambridge.

MacDonald, Sir, and Partners. 1987. *Kesem Irrigation Project Feasibility Study*. UNEP/FAO Report to WRDA, Volume 2, Annexes A–D.

Maffi, Luisa and Ellen Woodley. 2010. *Bio-cultural Diversity Conservation: A Global Sourcebook*, London: Earthscan.

Mammo, Mekbib, Girma Berhanu, and Taddese Yiberta. 1984. *The Nomadic Areas of Ethiopia Study Report*. Part 2 – The Physical Resources, Addis Ababa: United Nations Development Program, Eth/81/001, p. 71.

Manson, Katrina. 2014. Ethiopia uses electricity exports to drive ambition as an African power hub. *The Financial Times*, 16 February 2014. Available online at www.alloexpat.com/ethiopia_expat_forum/ethiopia-uses-electricity-exports-to-drive-ambition-as-an-a-t18916.html (accessed 10 July 2016).

Matonidi, Prosper B., Kjell Havnevik, and Atakilte Beyene (eds.). 2011. *Biofuels, Land Grabbing and Food Security in Africa*, London: Zed Books.

Mattravers, M. and M. Netsereab 1994. *The Senkelle Swayne's Hartebeest Sanctuary Management Plan*, EWCO, Mimeo, Addis Ababa; Cited in Nobuko Nishizaki. 2004. Resisting Imposed Wildlife Conservation: Arsi Oromo and the Senkelle Swayne's Hartebeest Sanctuary. *African Studies Monograph* 25, 2: 61–77.

McCann, J. C. 1997. The plough and the forest − narratives of deforestation in Ethiopia, 1840–1992. *Environmental History* 2, 138–59.

McCann, James. 1995. *People of the Plow: An Agricultural History of Ethiopia, 1800–1990*, Madison: University of Wisconsin Press.

McKee, Jonathan. 2007. *Ethiopia: Country Environmental Profile, 2007*. pp. 25–6. Available online at http://ec.europa.eu/development/icenter/repository/Ethiopia-ENVIRONMENTAL-PROFILE-08-2007_en.pdf (accessed 5 April 2011).

McKee, Jonathan. 2007. *Ethiopia: Country Environmental Profile*, p. 33. Available online at http://ec.europa.eu/development/icenter/repository/Ethiopia-ENVIRONMENTALPROFILE-08-2007_en.pdf (accessed 10 April 2011).

Mckee, Jonathan. 2007. *Ethiopia: Country Environmental Profile*, p. 40.

Mequanent, Dagnew, and Aboytu Sisay. 2015. Wetland Potential, Current Situation and its Threats in Tana Sub-Basin, Ethiopia. *World Journal of Environmental and Agricultural Sciences*, 1, 1, p. 10.

Merid, Fitsum. 2008. *National Nile Basin Water Quality Monitoring Baseline Report for Ethiopia*, Nile Basin Initiative, Trans-boundary Environmental Action Project, Addis Ababa, p. 15. Available online at http://nilerak.hatfieldgroup.com/English/NRAK/Resources/Document_centre/WQ_Baseline_report_Ethiopia.pdf (accessed 24 March 2011).

Mersha, Gebru, and Mwangi wa Gĩthĩnji. 2005. *Untying the Gordian Knot: The Question of Land Reform in Ethiopia*. Land, Poverty and Public Action Policy Paper No. 9, Institute of Social Studies/United Nations Development Program, New York, pp. 21–6; p. 23.

Metawie, A. F. 2004. History of Co-operation in the Nile Basin. *Water Resources Development* 20, 1: 47–63.

Meza, F. J., J. W. Hansen, and D. Osgood. 2008. Economic value of seasonal climate forecasts for agriculture: review of ex-ante assessments and recommendations for future research. *Journal of Applied Meteorology and Climatollgy*, 47, 1269–1286.

Michael, Yohannes Gebre, Saidou Magagi, Wolfgang Bayer, and Ann Waters-Bayer. 2011. *More than Climate Change: Pressures Leading to Innovation by Pastoralists in Ethiopia and Niger*. Paper presented at the International Conference on the Future of Pastoralism, 21–23 March, Organised by the Future Agricultures Consortium at the Institute of Development Studies, University of Sussex and Feinstein International Center of Tufts University. Available online at www.prolinnova.net/sites/default/files/documents/news/gebremichael_et_al.pdf (accessed 17 August 2011).

Miller, G. Tyler and Scott E. Spoolman. 2002. *Living in the Environment*. Belmont, CA: Wadsworth Group.

Mills, Greg. 2010. *Why Africa Is Poor and What Africans Can Do About It*, Johannesburg: Penguin, p. 170.

Minas, Getachew. 2001. Population dynamics and their impact on population growth and sustainable livelihood in Ethiopia. In: Food Security through Sustainable Land Use, *Proceedings of the Second National Workshop of NOVIB Partners on Sustainable Land Use*, Ababa, Addis, and Taye Assefa (ed.), 89–132; p. 91.

Ministry of Agriculture. 2014. ZemenawiGibrinatchen, vol. 3, No. 1, Yekatit 2006 E.C., p. 7.

Ministry of Health, 2005. *National Strategy for Child Survival in Ethiopia*, Addis Ababa: Family Health Department, Federal Ministry of Health, 2005. Available online at www.eshe.org.et/childsurvival/Child%20Survival%20Strategy.pdf (accessed 1 August 2011).

Ministry of Mines and Energy of Ethiopia (MoME). 2007. *The Biofuel Development and Utilization Strategy of Ethiopia*, Addis Ababa, Ethiopia.

Ministry of Natural Resources Development and Environmental Protection (MNRDEP). 1993. *Ethiopian Forestry Action Program*. Final Report, Volume II: The Challenge for Development, Addis Ababa, MNRDEP, p. 82.

Ministry of Water Resources (MoWR) and National Meteorological Services Agency (NMSA). 2001. Initial National Communication of Ethiopia to the United Nations.

Ministry of Water Resources (MoWR) and National Meteorological Services Agency (NMSA). 2004. Annual Climate Bulletin Year 2004, Addis Ababa: NMSA

Ministry of Water Resources (MoWR) and National Meteorological Services Agency (NMSA). 2001. *Initial National Communication of Ethiopia to the United Nations*

Framework Convention on Climate Change (UNFCCC), June, Addis Ababa, Ethiopia, p. 36.

Ministry of Water Resources (MoWR). 1998a. *Water Supply Development and Rehabilitation: Tariff and Asset Valuation Study*. Main Report, Addis Ababa: MoWR, p. 43.

Ministry of Water Resources. 1960. Omo-Gibe River Basin Integrated Development Master Plan Study. Final Report, Volume III, Development Zones and Areas.

Ministry of Water Resources. 2006. Five Year Irrigation Development Program: 2005/2006–2009/2010. Addis Ababa, Ethiopia: MoWR.

Moehlman, P. D., F. Kebede, and H. Yohannes. 1998. The African wild ass (*Equus africanus*): conservation status in the horn of Africa. *Applied Animal Behavior Science* 60(2, 3): 115–24.

Moehlman, P. D., H. Yohannes, R. Teclai, and Fanuel Kebede. 2008. Equus africanus. In: *IUCN Red List of Threatened Species*. Available online at http://en.wikipedia.org/wiki/ African Wild Ass (accessed 14 March 2011).

Moges, Kassaye Tekola. 2010. *Smallholder Farmers and Biofuel Farmers' Perspectives in Growing Castor Beans in Ethiopia*. Master's thesis, Uppsala, Swedish University of Agricultural Sciences, p. 8.

Mortimore, Michael. 1991. *A Review of Mixed Farming System in the Semi-Arid Zone of Sub-Saharan Africa*, Addis Ababa: International Livestock Center of Africa, p. 8.

MoWR. 1996e. *Tekeze River Basin Integrated Development Master Plan Project: Volume I*, Main Report, Addis Ababa: MoWR.

MoWR. 1997. *Abbay River Basin Integrated Development Project*. Volume V: Environment, Addis Ababa, MoWR, p. 54.

MoWR. 1997d. *Abbay River Basin Integrated Development Master Plan Project*, Volume V: Environment, Addis Ababa: MoWR, p. 41.

MoWR. 1998c. *Federal Water Policy and Strategy: Comprehensive and Integrated Water Resources Management*, Report, Vol. 1. Addis Ababa: MoWR, p. 15.

MoWR. 2002b. *Water Sector Development Program*. Final Report, Volume II, Addis Ababa: MoWR.

Mutia, Silas Mnyiri. 2009. Kenya's experience in managing climate change and water resource conflicts. In: *Climate Change and Trans-boundary Water Resources in Africa*, Workshop Report, 29–30 September, Mombasa, Kenya, p. 28.

Mwendera, E. J., M. A. Mohamed Saleem, and A. Dibabe. 1997. The effect of livestock grazing on surface runoff and soil erosion from sloping pasture lands in the Ethiopian highlands. *Australian Journal of Experimental Agriculture* 37: 421–30.

National Metreological Services Agency (NMSA). 2006. *Agro-meteorology Bulletin*, Addis Ababa: NMSA.

National Research Council. 2009. *Emerging Technologies to Benefit Farmers in Sub-Saharan Africa and South Asia*, Washington, DC: The National Academies Press, p. 39.

National Research Council. 2009. *Emerging Technologies to Benefit Farmers in Sub-Saharan Africa and South Asia*, Washington, DC: The National Academies Press, p. 212.

Native Solutions to Conservation Refugees. 2011. *The Take-Over of Ethiopia's Omo National Park by African Parks Foundation of the Netherlands*, pp. 1–4. Available online at www.danadeclaration.org/pdf/omotakeover.pdf (accessed 10 July 2016).

Ndulu, Benno J., Stephen A. O'Connell, Robert H. Bates, et al. (eds.). *The Political Economy of Growth in Africa: 1960–2000*, pp. 116–42; p. 119.

Newman, James. 1995. *The Peopling of Africa*. New Haven: Yale University Press.

Nievergelt, B., T. Good, and R. Guttinger. 1998. A survey of the Flora and Fauna of the Semien Mountains National Park, Ethiopia, Special Issue of *Walia*, Addis Ababa, Ethiopia.

Nigatu, Lisanework and Mesfin Tadesse. 1989. An ecological study of the vegetation of the Harenna Forest, Bale, Ethiopia. *SINET: Ethiopian Journal of Science* 12: 63–93.

Nissanke, Mchinko, and Erik Thorbecke (eds.). *The Poor under Globalization in Asia, Latin America, and Africa*, Oxford: Oxford University Press, pp. 368–97; p. 392.

Nkomo, J. C., A. O. Nyong, and K. Kulindwa. 2006. *The Impacts of Climate Change in Africa*. The Stern Review on the Economics of Climate Change, p. 26.

Nobre, Carlos A. 2009. How can current and future early warning systems be used to enhance adaptive capacity to climate change? *Development in a Changing World*, Blogs.WorldBank.org, October 5. Available online at https://blogs.worldbank.org/climatechange/howcan-current-and-future-early-warning-systems-be-used-enhance-adaptive-capacityclimate-change (accessed 27 October 2010).

Office of Environmental Health Hazard Assessment, State of California. 2009. *Microcystis: Toxic Blue-Green Algae*. Available online at http://oehha.ca.gov/ecotox/pdf/microfactsheet122408.pdf (accessed 23 December 2010).

Owen, Oliver S., and Daniel D. Chiras. 1990. *Natural Resource Conservation: An Ecological Approach*, New York: McMillan Publishing Co., pp. 59–77.

Padayachee, Vishu (ed.). *The Political Economy of Africa*, , pp. 85–109; See graph, p. 96.

Panayotou, Theodore. 1993. *Green Markets: The Economics of Sustainable Development*, San Francisco: Institute for Contemporary Studies, p. 3.

Pankhurst, Richard. 1962. The foundation and growth of Addis Ababa to 1935. *Ethiopian Observer* 6: 49–51.

Pankhurst, Richard. 1992. The history of deforestation and afforestation in Ethiopia prior to World War II. *Ethiopian Institute of Development Research* 2(2): 59–77.

Pankhurst, Richard. *Ethiopian Environmental History: Lions and Elephants in Ethiopian History II*, Addis Ababa, p. 7.

Pankhurst, Richard. Notes on the demographic history of Ethiopian towns and villages. *Ethiopia Observer* 9(1): 60–83.

Pearce, Fred. 2006. *When the Rivers Run Dry: Water – The Defining Crisis of the Twenty-First Century*, Boston: Beacon Press.

Pearce, Fred. 2012. *The Land Grabbers: The New Fight Over Who Owns the Earth*, Boston: Beacon Press.

Pearce, Fred. 2012. *The Land Grabbers: The New Fight Over Who Owns the Earth*, Boston: Beacon Press, pp. 5–6.

Peck, Emily. *Lake Turkana Conservation Science Program-WWF-US*. Available online at www.feow.org/ecoregion_details.php?eco=530 (accessed 7 December 2010).

Philips, Matt, and Jean-Bernard Carillet. 2006. *Ethiopia and Eritrea*, Third Edition, Guilford, CT: Lonely Planet, p. 211.

Philips, Matt, Jean-Bernard Carillet, and Zelalem Tefra Ashenafi. 1994. *Management Plan of Omo National Park*, Addis Ababa: Ethiopian Wildlife Conservation Organization, p. 11.

Plant Genetic Resources Center/Ethiopia (PGRC). 1986. *Ten Years of Collection, Conservation and Utilization*, 1976–1986, Addis Ababa: PGRC, p. 12.

Prabu, P.C. 2009. Impact of heavy metal contamination of Akaki River of Ethiopia on soil and metal toxicity on cultivated vegetable crops. *Electronic Journal of Environmental, Agricultural and Food Chemistry* 8, 9: 818–27. Available online at http://ejeafche.uvigo.es/component/option,com_docman/task,doc_view/gid,534/ (accessed 21 December 2010).

Rahmato, Dessalegn. 1998. Environmentalism and conservation in Wollo before the revolution. *Journal of Ethiopian Studies* 32: 43–86.

Rahmato, Dessalegn. 1999. Revisiting the Land Issue: options for change. *Economic Focus*, Volume 2, No. 4, August.

Rahmato, Dessalegn. 2009. *The Peasant and the State: Studies in Agrarian Change in Ethiopia 1950s–2000s*, Addis Ababa: Addis Ababa University Press.

Rahmato, Dessalgn. 1996. *Land and Agrarian Unrest in Wollo, Northeastern Ethiopia, Pre- and Post-Revolution*. IDR Research Report No. 46, Addis Ababa University.

Rahmato, Dessalgn. 1999. *Water Resource Development in Ethiopia: Issues of Sustainability and Participation*. Forum for Social Studies, Addis Ababa.

Rawlence, Ben. 2012. Why are Ethiopian eyes on Brussels? EuropeanVoice.com. Available online at www.europeanvoice.com/article/2012/july/why-are-ethiopianeyes-on-brussels-/74898.aspx (accessed 24 July 2012).

Red Cross Society.1987. *Pastoralist Rehabilitation and Development: Irrigation Projects in the Gewane and Assaita Areas*, Addis Ababa, p. 2.

Rice, Xan. 2010. Green concerns over Ethiopia's dam plans. *The Sydney Morning Star*, 12 April.

Richardson, Paul. 2012. Ethiopia to accelerate land commercialization amid opposition. Available online at www.businessweek.com/news/2012-03-23/ethiopia-toaccelerate-land-commercialization-amid-opposition (accessed 6 April 2012).

Riché, Béatrice, Excellent Hachileka, Cynthia B. Awuor, and Anne Hammill. 2009. Cited in *Climate-Related Vulnerability and Adaptive-Capacity in Ethiopia's Borana and Somali Communities*. Final assessment report. Save the Children (UK). Available online at www.iisd.org/pdf/2010/climate_ethiopia_communities.pdf, p. 22 (accessed 4 November 2010).

Ridgewell, Andrew, Getachew Mamo, and Fiona Flintan (eds.). 2007. *Gender & Pastoralism*, Volume 1: Rangeland & Resource Management in Ethiopia, Addis Ababa, p. 49. Available online at www.sahel.org.uk/pdf/Gender%20&%20Pastoralism%20Vol%201%20-%20ebook.pdf. (accessed 12 November 2010).

Rosen, Armin. 2012. The Zenawi paradox: the Ethiopian leader's good and terrible legacy. *The Atlantic*, July 21. Available online at www.theatlantic.com/international/archive/2012/07/the-zenawi-paradox-an-ethiopian-leaders-good-andterrible-legacy/260099/ (accessed 21 July 2012).

Rosillo-Calle, Frank, and Francis X. Johnson (eds.). *Food versus Fuel: An Informed Introduction to Biofuels*, London: Zed Press, pp. 138–63; pp. 153–4.

Said, Ali. 1994. *Pastoralism and State Policies in Mid-Awash Valley: The Case of the Afar, Ethiopia, African Arid Lands*. Working Paper Series No. 1.

Save the Children (UK). 2009. *Climate-Related Vulnerability and Adaptive-Capacity in Ethiopia's Borana and Somali Communities*. Final assessment report, p. 12. Available online at www.iisd.org/pdf/2010/climate_ethiopia_communities.pdf (accessed 4 November 2010).

Shackleton, D. M. 1997. *Wild Sheep and the Relatives*. IUCN/SSC Caprinae Specialist Group, IUCN, Gland, Switzerland and Cambridge, UK.

Shinn, David H. and Thomas P. Ofcansky. 2004. *Historical Dictionary of Ethiopia*, Oxford: The Scarecrow Press.

Simone, Abdou Maliq. 1994. *From the City Yet to Come: Changing African Life in Four Cities*. Durham: Duke University Press.

Singh, Harjinder. 1987. *Agricultural Problems in Ethiopia*, Delhi: Gian Publishing House.

Sissay, Lemlem. 2003. Biodiversity potentials and threats to the southern Rift Valley lakes of Ethiopia. In: *Wetland of Ethiopia: Proceeding of a Seminar on the Resources and Status of Ethiopia's Wetlands*, Yilma Dellelegn and K. Geheb (eds.), IUCN, Gland, p. 116.

Skinner, Djihan. 2010. *Rangeland Management for Improved Pastoralist Livelihoods: The Borana of Southern Ethiopia*. MA thesis in Development and Emergency Practice, Oxford

Brookes University, April, p. 10. Available online at www.brookes.ac.uk/schools/be/ oisd/architecture/cendep/dissertations/DjihanSkinner.pdf (accessed 4 November 2010).

Soromessa, Teshome, Demel Teketay, and Sebsebe Demissew. 2004. Ecological study of the vegetation in Gamo Gofa zone, southern Ethiopia. *Tropical Ecology* 45(2): 209–21.

Stattersfield, A. J., M. J. Crosby, A. J. Long, and D. C. Wege. 1998. *Endemic Bird Areas of the World: Priorities for Conservation*, BirdLife International, Cambridge, UK.

Stern, N. 2007. *Stern Review on the Economics of Climate Change*, London: Cambridge University Press. Available online at www.hm-treasurey.gov.uk/independent_reviews/ stern_review_economics_climate_change/stern_review_report.cfm.

Sutcliffe, J. P. 1995. Soil conservation and land tenure in highland Ethiopia. Special issue on Land Rights and Access to land in Post-Derg Ethiopia. *Ethiopian Journal of Development Research* 17: 63–80.

Swain, A. 1997. Ethiopia, the Sudan and Egypt: the Nile River dispute. *The Journal of Modern African Studies*, 35(4): 675–94.

Szirmai, Adam. 2005. *The Dynamics of Socio-Economic Development*, Cambridge: Cambridge University Press.

Tadesse, Zenbework (ed.). *Proceedings of the Symposium of the Forum for Social Studies*, Addis Ababa, 15–16 September, pp. 92–8.

Tato, Kebede. 1990. Ethiopian soil conservation program and its future trends. In: *National Conservation Strategy Conference Document*, Vol. 2, Introduction and Africa's Experience of National Conservation Strategies, ONCCP, Addis Ababa.

Tedla, Shibru. 1995. Protected Areas Management Crisis in Ethiopia. *Walia*16, pp. 17–30; p. 20.

Tefera, B.and A. Bekele (2004). Population status and structure of mountain nyala in the Bale Mountains National Park, Ethiopia. *African Journal of Ecology* 42(1): 1–7.

Tefera, Bezuayehu and Geert Sterk. 2008. Hydropower-Induced Land Use Change in Fincha'a Watershed, Western Ethiopia: Analysis and Impacts, *Mountain Research and Development* 28(1): 72–80.

Teferi, Abate. 1995. Land redistribution and inter-household relations: the case of two communities in northern Ethiopia. *Ethiopian Journal of Development Research* 17: 23–40.

Tegene, Belay. 1998. Indigenous soil knowledge and fertility management practices of the south Wollo Highlands. *Journal of Ethiopian Studies* 32:123–58.

Teketay, Demel. 1992. Human impact on a natural montane forest in southeastern Ethiopia. *Mountain Research Development* 1212: 393–400.

Teketay, Demel. 1995. Floristic composition of Dakata Valley, Southeast Ethiopia: an implication for the conservation of biodiversity. *Mountain Research and Development* 15(2): 183–6.

Teketay, Demel. 1999. Agroforestry: prospects and adoption by Ethiopian farmers. *Walia* 20: 28–50.

Teketay, Demel. 1999. *Deforestation, Wood Famine, and Environmental Degradation in Ethiopia's Highland Ecosystems: Urgent Need for Action*, Forest Stewardship Council (FSC Africa), Kumasi, Ghana, pp. 61–9; pp. 72–5.

Teketay, Demel. 2000. Facts and experiences on Eucalypts in Ethiopia and elsewhere: ground for making wise and informed decision. *Walia* 20: 25–46.

Teshale, Berhanu, Ralph Lee, and Girma Zawdie. 2002. Development initiatives and challenges for sustainable resource management and livelihood in the Lake Tana region of northern Ethiopia. *International Journal of Technology Management and Sustainable Development* 1(2): 111–24.

The International Research Institute for Climate and Society, IRI Scientist Wins NSF CAREER Award, October 15, 2010. Available online at http://portal.iri.columbia.edu/portal/server.pt?open=512&objID=699&parentname=CommunityPage&parentid=0&mode=2&in_hi_userid=2&cached=true (accessed 27 October 2010).

The Oakland Institute. 2011. *Understanding Land Investment Deals in Africa, Country Report: Ethiopia*, Oakland: The Oakland Institute.

The World Bank. 2006. *Managing Water Resources to Maximize Sustainable Growth*, Ethiopia, Report 36000-ET, World Bank: Washington, DC, pp. 21–2.

The World Bank. 2007. *Rising Global Interest in Farmland: Can It Yield Sustainable and Equitable Benefits?* Washington, DC: The World Bank, p. 44.

Thiene, Michele L., Robin Abell, Melanie L. J. Stiassny, and Paul Skelton. 2005. *Freshwater Ecoregions in Africa and Madagascar: A Conservation Assessment*, Washington, DC: World Wildlife Fund, p. 353.

Thompson, Larry. 2005. Ethiopia: Local people burned out of homes to make way for national park, *Refugees International*, Washington, DC, April.

Tolossa, Degefa and Axel Baudouin. 2004. Access to natural resources and conflicts between farmers and agro-pastoralists in Borkena Wetland, Northeastern Ethiopia. *Norwegian Journal of Geography* 58(3): 97–112.

Toulmin, Camilla. 2009. *Climate Change in Africa*, London: Zed Press.

Transitional Government of Ethiopia (TGE). 1994. *A Review of the Water Resources of Ethiopia*, Addis Ababa, p. 79.

Troccoli, A., M. Harrison, D. L. T. Anderson, and S. J. Mason (eds). 2008. *Seasonal Climate: Forecasting and Managing Risk*. Dordrecht: Springer, pp. 351–95. doi: 10.1007/978-1-4020-6992-5-13.

Tsehaye, Addis. 1986. Utilization of forest products in Ethiopia. *Sinet: Ethiopian Journal of Science* (suppl.): 45–60.

Turnton, D. 1987. The Mursi and National Park development in the lower Omo Valley. In: *Conservation in Africa: People, Policies and Practices*, Anderson, D. M., and R. H. Grove (eds.), Cambridge: Cambridge University Press, pp. 169–86.

UN Office for the Coordination of Humanitarian Affairs. 2006. *Ethiopia: Dire Dawa Floods*, Ref. OCHA/GVA-2006/0143.

UNEP. 2009. *Semien Mountains National Park*, New York: UNEP.

UNESCO and EWCO. 1986. *Management Plan: Semien Mountains National Park and Surrounding Rural Areas*, Addis Ababa, Ministry of Agriculture, p. 56.

UNESCO World Heritage Committee and the Ethiopian Wildlife Conservation Organization (EWCO). 1986. *Management Plan: Semien Mountains National Park and Surrounding Rural Areas*, Addis Ababa: Ministry of Agriculture, p. 10.

UNICEF, Ethiopia: basic indicators. Available online at www.unicef.org/infobycountry/ethiopia_statistics.html (accessed 21 March 2013).

United Nations Conference and Trade and Development (UNCTAD). 2012. *Economic Development in Africa: Report 2012: Structural Transformation and Sustainable Development*, New York and Geneva: UNCTAD, p. 123.

United Nations Development Program (UNDP). 2013. The rise of the South: Human progress in a diverse world. *Human Development Report 2013*, New York: UN, p. 196.

United Nations Development Program Emergencies Unit for Ethiopia (UNDP – EUE). 1988. *South East Rangelands Project (SERP)*, Addis Ababa: UNDP.

United Nations Development Program. 2013. *Human Development Report 2013*, New York: UNDP, Table 13, p. 192.

United Nations Development Program. 2013. *Human Development Report* 2013, New York: UNDP, Table 13, p. 192.

United Nations Economic Commission for Africa. 2000. *Trans-boundary River/Lake Basin Water Development in Africa: Prospects, Problems, and Achievements*, Addis Ababa: UNECA. p. 28.

United Nations Educational, Scientific, and Cultural Organization World Water Assessment Program. 2004. *National Water Development Report for Ethiopia*, Final Report, Addis Ababa, Ministry of Water Resources, December, UN-WATER/WWAP/2006/7, p. 71. Available online at http://unesdoc.unesco.org/images/0014/001459/145926e.pdf (accessed 26 November 2010).

United Nations Educational, Scientific, and Cultural Organization (UNESCO). 2004. *National Water Development Report for Ethiopia*, Addis Ababa: World Water Assessment Program, p. 227. Available online at http://unesdoc.unesco.org/images/0014/001459/145926e.pdf (accessed 2 December 2010).

United Nations Educational, Scientific, and Cultural Organization (UNESCO). 2004. *National Water Development Report for Ethiopia*. Final Report, Addis Ababa, MoWR, December, UN-WATER/WWAP/2006/7. Available online at http://unesdoc.unesco .org/images/0014/001459/145926e.pdf (accessed 26 November 2010).

United Nations Educational, Scientific, and Cultural Organization, World Water Assessment Program. 2004. *National Water Development Report for Ethiopia*. Final Report, Addis Ababa, Ministry of Water Resources, December 2004, UN-WATER/ WWAP/2006/7, EWNHS, p. 111. Available online at http://unesdoc.unesco.org/ images/0014/001459/145926e.pdf, (accessed 26 November 2010).

United Nations Environmental Program (UNEP). 2009. *Semien Mountains National Park*, UNEP: New York, 2009.

United Nations. 2012. COMTRADE.

United States Population Reference Bureau. *2011. World Population Data Sheet*. Available online at www.prb.org/DataFinder/Geography/Data.aspx?loc=278 (accessed 21 March 2013).

Unruh, John D. 2006. Changing conflict resolution institutions in the Ethiopian pastoral commons: the role of armed confrontation in rule-making. *GeoJournal* 64(Spring): 225–37; p. 230.

Urban, Emil K. 1969. Ecology of water birds of four Rift Valley lakes in Ethiopia. *Ostrich* 40 (Suppl. 1): 315–22; p. 320.

Urban, Emil K. 1970. Ecology of water birds of four Rift Valley lakes in Ethiopia. *Ostrich Sup.*, 8: 315–21; p. 318.

US Central Intelligence Agency, The World Factbook. Available online at www.cia.gov/ library/publications/the-world-factbook/geos/et.html (accessed 21 March 2013).

Vaughan, Jenny. 2011. Ethiopia land lease risks displacement, *AFP*, July 29, 2011. Available online at www.google.com/hostednews/afp/article/ALeqM5hURPr3NAF STH794ZxOHq3wYAuYfA?docId=CNG.1046baaee64b2f1d60afa5a979a5d3a0.91 (accessed 5 August 2011).

Vidal, John. 2010. Billionaires and Mega-Corporations Behind Immense Land Grab in Africa. *Mail and Guardian*, 10 March, p. 5.

Water Resources Development Authority, Transitional Government of Ethiopia. 1995. *Survey and Analysis of the Upper Baro-Akobo Basin*. Final Report, Volume I – Main Report, Addis Ababa: Addis Resources Development-GEOSERV (ARDCOGEOSERV), pp. 24 and 31.

Water Resources Development Authority. 1989. *Amibara Irrigation Project II Pastoralists and Forestry Development Studies*. Final Report, p. 2.3.

Waterbury, John. 1979. *The Hydropolitics of the Nile Valley*, Syracuse: Syracuse University Press.

Waterbury, John. 1987. Legal and Institutional Arrangements for Managing Water Resources in the Nile Basin. *Water Resources Development* 3: 92–103.

Waterbury, John. 2002. *The Nile Basin: National Determinants of Collective Action*, New Haven: Yale University Press, p. 16.

Watershed, Ethiopia: Land Use Changes, Erosion Problems, and Soil and Water Conservation Adoption, Wageningen: Wageningen University, The Netherlands. No date. Available online at www.mekonginfo.org/mrc_en/doclib.nsf/0/e1dfbbefb9263e6b4725724a00123f75/$file/07_abstr_sterk_tefera.pdf (accessed 5 January 2011).

Whittington, Dale and McClelland, Elizabeth M. 1992. Opportunities for regional and international cooperation in the Nile Basin. *Water International* 17: 144–54.

Whittington, Dale Xun Wu, and Claudia Sadoff. 2005. Water resources management in the Nile basin. The economic value of cooperation. *Water Policy* 7: 227–52.

Winegardner, Duane L. 2008. The fundamental concept of soil. In: *Environment: An Interdisciplinary Anthology*, Adelson, Glenn, James Engell, Brent Ranalli, and K. P. Van Anglen (eds.), New Haven: Yale University Press, pp. 415–8; p. 416.

Wolde Selassie, Gebremarkos. 1998. The forest resources of Ethiopia past and present. *Walia* 19, 10–28.

Wolde-Ab, Asnakew. 1987. Physical properties of Ethiopian vertisols. In: *Proceedings of a Conference on Management of Verisols in Sub-Saharan Africa held at ILCA*, Addis Ababa, 31 August–4 Septemeber 1987, pp. 111–23.

Woldetsadik, Muluneh. 2003. Population growth and environmental recovery: more people, more trees, lesson learned from West Gurageland. *Ethiopian Journal of the Social Sciences and Humanities*, 1(1): 1–33.

Woldu, Zerihun and Mesfin Tadesse. 1990. The status of vegetation in the Lakes region of the Rift Valley of Ethiopia and the possibilities of its recover. *SINET: Ethiopian Journal of Science* 3: 97–120.

Wood, Adrian P. 1993. Natural resource conflicts in southwest Ethiopia: state, communities, and the role of the National Conservation Strategy in the search for sustainable development. *Nordic Journal of African Studies* 2(2): 83–99.

Wood, Adrian. 2000. Wetlands, gender, and poverty: some elements in the development of sustainable and equitable wetland management. In: *Proceedings of a Seminar on the Resources and Status of Ethiopia's Wetlands*, Abebe, Yilma D., and Kim Geheb (eds.), The World Conservation Union (IUCN).

Wood, R. B., M. V. Prosser, and F. M. Baxter. 1978. Optical Characteristics of the Rift Valley Lakes, Ethiopia. *SINET: Ethiopian Journal of Science* 1(2): 73–85.

Woody Biomass Inventory and Strategic Planning Project (WBISPP). 1995. *Annex 3: Socio-Cultural and Economic Aspects of Crop, Livestock and Tree Production*, WBISPP, November 30, p. 23.

Worede, Melaku, Tesfaye Tesemma, and Regassa Feyissa. 1999. Ch. 6: Keeping diversity alive: An Ethiopian perspective. In: *Genes in the Field: On-Farm Conservation of Crop Diversity*, Stephen B. Brush (ed.), CRC Press, pp. 143–61.

World Bank. 2006. *Managing Water Resources to Maximize Sustainable Growth, Ethiopia*, Report 36000-ET, World Bank: Washington, p. 28.

World Bank. 2012. Inclusive green growth, *World Bank Research Digest*, 6, 4 (Summer), pp. 4–5.

World Commission on Environment and Development. 1987. *Our Common Future.* Oxford: Oxford University Press, p. 52.

World Meteorological Organization. 2003. *The Associated Program on Flood Management Integrated Flood Management: Case of Ethiopia.* Integrated Flood Management (by Kefyalew Achamyeleh). Technical Support Unit (ed.). Available online at www.apfm. info/pdf/case_studies/cs_ethiopia.pdf (accessed 11 March 2011).

Wylde, Augustus B. 1901. *Modern Abyssinia*, London: Methuen, p. 449.

Yeshitela, Kumelachew and Tamrat Bekele. 2002. Plant community analysis and ecology of Afromontane and transitional rainforest vegetation of southwestern Ethiopia. *SINET: Ethiopian Journal of Science* 25: 155–75.

Yilma, Sileshi and G. R. Demaree. 1995. Rainfall variability in the Ethiopian and Eritrean highlands and their link with the Southern Oscillation Index. *Journal of Biogeography* 22: 945–52.

Yilma, Sileshi and U. Zanke. 2004. Recent change in rainfall and rainy days in Ethiopia. *International Journal of Climatology* 24: 973–83.

Yilma, Yeraswork. 1997. Integrated Pest Management (IPM) Experience in the Amhara Region. In: *Proceedings of Integrated Pest Management Workshop in the Amhara Region*, 24–26 February, Dessie, 28–33.

Zenebe, Wudineh. 2007. Ethiopia's New Energy Strategy Opens Way for Biofuel. *Addis Fortune*, 4 November. Available online at http://akababi.org/bionov07041.htm (accessed 27 February 2012).

Zewde, Bahru and Siegfried Pausewang (eds.). 2002. *Ethiopia: The Challenge of Democracy From Below*, Addis Ababa: Forum for Social Studies.

Zewde, Bahru. 1998. Forests and forest management in Wollo in historical perspective. *Journal of Ethiopian Studies* 32: 87–121.

Zinabu, Gebre Mariam and D. Elias, D. 1989. Water resources and fisheries management in the Ethiopian rift valley lakes. *SINET: Ethiopian Journal of Science* 12:95–109.

Index

Aba Samuel, Lake 127, 141
Abaya, Lake 17, 25, 124, 128,
  135–136, 138
Abijata, Lake 124, 128–133, 138
Abijata-Shala Lakes National Park 130,
  157, 167–168
Adaptation 36, 186
Afforestation 7, 40, 80, 82–83, 88, 100
African Parks Network (APN) 171–172
Agroforestry 7, 22, 38–39, 72, 87–90, 95,
  97, 100, 184
Agropastoral 52, 54, 61, 63, 66–72, 117,
  120, 171
Akaki River 127
Akobo River 24, 70, 96, 120, 142, 172–173
Al-Sisi, President 123
Al Wabra Wildlife Preservation
  (AWWP) 174
Anferara-Wadera Forests 92
Arable land 5, 8, 15–16, 19, 22, 24, 26, 28,
  36–37, 40, 101
Arithmetic density 5
Aswan High Dam 123
Awash National Park 56, 58, 72, 154–159
Awash River 17, 19, 24, 31, 33, 53–56,
  112, 116–117, 127–128, 133, 142–143,
  145, 147, 150–151, 160
Awash Valley Authority (AVA) 55

Babille Elephant Sanctuary 174–175
Bahir-el-Azraq 114
Bale Mountains National Park (BMNP)
  90–91, 157, 164–167
Baro River 17, 68, 70–71, 96, 112, 120,
  143, 172–173
Bega 32
Belg 32
Biochemical oxygen demand (BOD) 135
Biodiversity 11, 19, 43, 93–94, 96, 103

Biofuels 26, 30–31, 42
Biological diversity 5, 11
Biomass 30–31
Biosphere Reserves 89, 95–96
Birds 42–43, 78, 90, 96, 98, 130–131,
  133–134, 138, 141, 153–154, 160,
  167–169, 172, 174, 176
Birkedas 65, 67
Blue Nile River 10, 25, 36, 112–115,
  122–125, 136, 147, 150
Bonga Forests 94–95
Bulbulla River 131
Burundi 123

Chamo, Lake 124, 128, 135–138, 168
Chemical oxygen demand (COD) 135
Chew Bahir, Lake 58, 124, 128,
  135–137, 170
Chew Bahir Wildlife Reserve 137
Child mortality 3, 5
Climate change 9, 31–36, 43, 61, 72
Community-based conservation 159
Contour farming 38
Conventional high-input systems 39–40
Crop residues 22–23, 37, 66
Crop rotation 3, 14, 37–38

Danakil Depression 36
Dawa River 59, 61, 112, 118, 122
Declaration of Principles (DoPs) 123
Deforestation 4, 6,9, 30–31, 35–36,
  79–82, 86, 92, 95–96, 99, 101, 113,
  155, 168, 184
Democratic Republic of Congo 2, 120
Derg regime 25, 82–85, 139–140, 158, 169
Desertification 4
Desse'a State Forest 96–97
Disaster Prevention and Preparedness
  Authority 36

Dissolved oxygen (DO) 135
Djibouti 10, 63–64, 122, 125
Drip irrigation 26
Drought 19, 24, 32–36, 42–43, 52–54, 56, 58, 60–64, 67, 72, 79, 82, 97, 112, 114, 117, 124, 127–134, 134, 137, 144–145, 151, 161, 184

Early warning systems 35
Ecological degradation 2, 6, 10, 81
Egypt 115, 122–124, 134, 144, 146–147, 185
El Nino 35
El Nino/Southern Oscillation (ENSO) 35
Environmental conservation 3
Environmental degradation 6
Environmental rehabilitation 3, 83–84
Eritrea 10, 160
Ethiopian Environmental Protection Authority 86
Ethiopian Forestry Action Plan (EFAP) 86
Ethiopian Heritage Trust (EHT) 90
Ethiopian People's Revolutionary Democratic Front (EPRDF) 85
Ethiopian Plant Genetic Resources Center (PGRC/E) 103
Ethiopian Wildlife Conservation Authority (EWCA) 177
Ethiopian Wildlife Conservation Organization (EWCO) 155, 159, 162, 174
Ethiopian Wildlife and Natural History Society (EWNHS) 90, 93, 140
Ethiopian wolf 162, 165–166
Eucalyptus 39, 80–83, 100–101
Eutrophication 126, 130, 135

Fallowing 38
Famine 4, 34–35, 56, 61, 82, 94, 13 9
Famine Early Warning System Network 34
Farming systems
    Enset-root farming 20, 22–23
    Large-scale farming 28–29
    Mixed cereal-livestock farming 20–22
    Shifting cultivation 24
Fertility rate 5
Fishing 5, 71, 119, 120, 126–127, 129, 138
Flooding 34–35, 42–43, 54–55, 65, 70, 116, 118–119, 127, 134, 184
Floriculture 29–30
Forest conservation 80, 86, 102

Forest management 82, 86, 95, 102–103, 185
Forest User Society (FUS) 103
Fuelwood 2, 6, 30, 38–39, 81–82, 84, 99–100, 102, 121, 157, 160

Gelada baboon 162–163
Gambela National Park 96, 143, 157, 172–173
Genale River 58–59, 92, 112, 118, 122
Gender (*see also* women) 31
Genetic diversity 8, 19 *see also* biological diversity
Gibe River 118, 122
Global Assessment of Soil Degradation (GLASOD) 36
Global Energy Pacific 31
Grand Ethiopian Renaissance Dam (GERD) 123
Greenhouse gases 32
Groundwater 8, 10, 24, 31–32, 65, 112–114, 130–131, 133, 136, 138–139
Gulf States 27

Habitats 4, 8, 11, 19, 28, 31, 36, 39, 78, 89, 92, 99, 114, 125–127, 134–138, 141–142, 153–155, 157, 160, 162–166, 168, 174–176
Haile Selassie, Emperor 55, 57, 81
Harenna Forests 90–91
Hawassa, Lake 124, 128, 133–135, 141
High-input system 39

Industrialisation 41, 136, 143, 185
Infant mortality 5
Institute of Biodiversity Conservation and Research (IBC) 103
Integrated production system 38
Intercropping 38
International Union for the Conservation of Nature (IUCN) 163, 165, 177
Irrigation 2, 3, 9–10, 14, 16–19, 23–28, 31, 40–41, 43, 55–57, 64, 67, 72, 82–83, 96, 114, 117, 128–131, 134, 137, 185

Juba River 117–118

Kaffa Forests 94
Kenya 10, 58, 61, 63, 118–119, 122–123–124, 185
Khartoum 114, 123
Kiremt 32

Koka Dam 55,
Koka, Lake 127–128

Land degradation 6–7, 36–37, 61,
    78, 184
Land reform 3, 41
Land tenure systems 2, 4, 40–41, 54
Langano, Lake 88, 124, 128, 131,
    133, 138
Low-input farming 39

Mago National Park 71, 170–171
Mehr 32
Menagesha State Forest 89
Menelik II, Emperor 80, 154, 175
Millennium Development Indicators 3
Mille-Serdo Wild Ass Reserve 160
Mono-cropping 28
Mortality rate 4
Munessa-Shashamane State Forest 88

National Forest Priority Areas (NFPAs)
    82, 86–88, 92–94, 103
National Meteorological Agency of
    Ethiopia 34
Nationalization 3
Nature and Biodiversity Conservation
    Union (NABU) 96
Nech Sar National Park 136, 157,
    168–170, 172
Nile Basin Initiative (NBI) 122
Non-renewable resources 6
Non-timber forest products (NTFPs)
    93, 95, 103
Non-traditional crops 29

Omo National Park 119, 157, 170–171
Omo River 10, 17, 68–72, 112, 118–119,
    145, 170
Overgrazing 6, 9, 55, 60–61, 63, 66, 72

Participatory forest management groups
    (PFMGs) 95, 102
Pibor River 120, 142
Population density 5, 23
Population growth 4–6, 10
Poverty 1–2, 6–7, 25, 41, 122
Private ranching 62

Rangelands 4, 24, 52, 55, 58, 60–64, 66,
    71–73, 184, 186
Ras Dashen 34, 153
Re-afforestation 83, 87, 89–90
Reforestation 7, 41, 84, 100

Regional Forest Priority Areas (RFPAs)
    88, 90, 92, 94
Relief Society of Tigray (REST) 97
Renewable resources 6
Rwanda 123

Salinisation 4, 8, 10, 19, 124
Saudi Arabia 27, 67
Sea surface temperature (SST), 35
Sedimentation 128, 136
Semien Mountains National Parks
    155–157, 161–164
Senkelle Swayne's Hartebeest
    Sanctuary 174
Shala, Lake 124, 128, 130–133, 138
Sodium bicarbonate 131
Soil conservation 7, 40, 82–83, 143
Soil degradation 2, 19, 37
Soil erosion 2, 4, 6, 8, 22, 33, 36–38, 43,
    81–82, 90, 184–185
Soil types
    Acrisols 18
    Cambisols 15
    Fluvisols 17
    Lithosols 18
    Luvisols 18
    Mollisols 17–18
    Nitosols 15
    Regosols 18
    Solonchaks 18
    Vertisols 15–16
    Xerosols 18
    Yermosols 18
Somalia 10, 59, 61, 63–64, 67–68,
    117–118, 122, 160, 185
Somaliland 64
South Korea 27
South Sudan 142
Sudan 10, 114–115, 119–120,
    124, 173, 185
Sudan People's Liberation Army
    (SPLA) 173
Sustainability 9, 11, 24, 86, 96–97,
    100, 177
Sustainable development 7–8, 25
Sustainable Development and
    Poverty Reduction Program
    (SDPRP), 25
Sustainable pastoralism 57

Tama Wildlife Reserve 71
Tana, Lake 114, 123, 125–127, 138,
    141, 168
Tanzania 123

Tekeze River 10, 112, 115, 144
Temperate 1, 17, 21, 78
Topography 14–16, 58, 64, 78, 92, 113, 120, 172
Turkana, Lake 17, 118–120, 124, 145–147

Uganda 123
Under-five mortality 5
United Nations Development Programme (UNDP) 34
United Nations Educational, Scientific and Cultural Organization (UNESCO) 72, 89, 95–96, 155, 162, 171
United States Agency for International Development (USAID) 155
Urban 3, 6, 8, 10, 24, 42, 52, 58, 63, 80, 82–83, 87, 94–96, 99–101, 112, 121, 126, 154, 166
Urbanisation 81, 121, 129, 134, 136, 141, 143, 185

Wabi Shebelle River 10, 16–17, 19, 25, 36, 112, 117, 157, 165, 175
Walia ibex 162–163, 166
Water pollution 4, 10–11, 113, 126–128, 136–137, 143, 185
Watersheds 10, 37, 81, 88–89, 95, 127–128, 140, 142–143
Wetlands 28, 31, 125–127, 135–143, 172, 185
Women (*see also* gender) 21, 28–29, 31, 63, 84, 99, 140
World Commission on Environment and Development 7
World Food Program 27, 82–83
World Wildlife Fund 155

Yabelo Game Sanctuary 176
Yangudi-Rassa National Park 160
Yewof Washa Forests 98

Ziway, Lake 124, 128–131, 133, 138

Printed and bound by CPI Group (UK) Ltd, Croydon, CR0 4YY

21/10/2024

01777089-0004